Springer Theses

Recognizing Outstanding Ph.D. Research

AF896098

Aims and Scope

The series "Springer Theses" brings together a selection of the very best Ph.D. theses from around the world and across the physical sciences. Nominated and endorsed by two recognized specialists, each published volume has been selected for its scientific excellence and the high impact of its contents for the pertinent field of research. For greater accessibility to non-specialists, the published versions include an extended introduction, as well as a foreword by the student's supervisor explaining the special relevance of the work for the field. As a whole, the series will provide a valuable resource both for newcomers to the research fields described, and for other scientists seeking detailed background information on special questions. Finally, it provides an accredited documentation of the valuable contributions made by today's younger generation of scientists.

Theses may be nominated for publication in this series by heads of department at internationally leading universities or institutes and should fulfill all of the following criteria

- They must be written in good English.
- The topic should fall within the confines of Chemistry, Physics, Earth Sciences, Engineering and related interdisciplinary fields such as Materials, Nanoscience, Chemical Engineering, Complex Systems and Biophysics.
- The work reported in the thesis must represent a significant scientific advance.
- If the thesis includes previously published material, permission to reproduce this must be gained from the respective copyright holder (a maximum 30% of the thesis should be a verbatim reproduction from the author's previous publications).
- They must have been examined and passed during the 12 months prior to nomination.
- Each thesis should include a foreword by the supervisor outlining the significance of its content.
- The theses should have a clearly defined structure including an introduction accessible to new PhD students and scientists not expert in the relevant field.

Indexed by zbMATH.

Christina Hofer

Detection Efficiency and Bandwidth Optimized Electro-Optic Sampling of Mid-Infrared Waves

Doctoral Thesis accepted by
Ludwig Maximilians University Munich, Garching,
Germany

Springer

Author
Dr. Christina Hofer
Max Planck Institute of Quantum Optics
Garching, Bayern, Germany

Ludwig Maximilian University of Munich
Munich, Bayern, Germany

Supervisor
Prof. Ferenc Krausz
Max Planck Institute of Quantum Optics
Garching, Bayern, Germany

Ludwig Maximilian University of Munich
Munich, Bayern, Germany

Center for Molecular Fingerprinting
Molekuláris-Ujjlenyomat Kutató
Közhasznú Nonprofit Kft.
Budapest, Hungary

ISSN 2190-5053 ISSN 2190-5061 (electronic)
Springer Theses
ISBN 978-3-031-15330-3 ISBN 978-3-031-15328-0 (eBook)
https://doi.org/10.1007/978-3-031-15328-0

© The Editor(s) (if applicable) and The Author(s), under exclusive license to Springer Nature Switzerland AG 2022
This work is subject to copyright. All rights are solely and exclusively licensed by the Publisher, whether the whole or part of the material is concerned, specifically the rights of translation, reprinting, reuse of illustrations, recitation, broadcasting, reproduction on microfilms or in any other physical way, and transmission or information storage and retrieval, electronic adaptation, computer software, or by similar or dissimilar methodology now known or hereafter developed.
The use of general descriptive names, registered names, trademarks, service marks, etc. in this publication does not imply, even in the absence of a specific statement, that such names are exempt from the relevant protective laws and regulations and therefore free for general use.
The publisher, the authors, and the editors are safe to assume that the advice and information in this book are believed to be true and accurate at the date of publication. Neither the publisher nor the authors or the editors give a warranty, expressed or implied, with respect to the material contained herein or for any errors or omissions that may have been made. The publisher remains neutral with regard to jurisdictional claims in published maps and institutional affiliations.

This Springer imprint is published by the registered company Springer Nature Switzerland AG
The registered company address is: Gewerbestrasse 11, 6330 Cham, Switzerland

Supervisor's Foreword

Electric field oscillations are now experimentally accessible in the THz-to-PHz frequency range. Their measurement—if performed with high sensitivity—can deliver the most comprehensive information attainable by optical spectroscopy. Yet, nonlinear mixing underlying optical field sampling imposes a trade-off between bandwidth and efficiency and compromises the sensitivity of field-resolved spectroscopy.

Christina Hofer's thesis has addressed optimization strategies of this trade-off and achieved sensitivities close to the ultimate limit of capturing all photons within the temporal gate. In addition to these efforts at the forefront of ultrafast metrology, the dissertation describes the application of the newly developed techniques to field-resolved infrared (IR) vibrational spectroscopy of real, biomedical samples.

The past decade has spawn powerful, highly phase-stable, broadband IR light pulses at multi-MHz repetition rates, with brilliance values exceeding even those of IR beamlines at synchrotron facilities. The coherent response of molecular samples to this temporally confined IR excitation can be isolated and sensitively detected by suitable time-resolved techniques. The goal of this thesis work was the design, construction and application of *electro-optic sampling* (EOS) of IR waves for highly sensitive, high-detection-dynamic-range field-resolved spectroscopy in the molecular fingerprint region.

EOS of mid-IR waves can be understood in the framework of a two-step model. In the first step, mid-IR photons within the temporal gate defined by the electric field of the gate pulse are upconverted, e.g., by sum-frequency generation. In the second step, the upconverted field is coherently superimposed on the gate field and balanced heterodyne detection delivers a signal containing amplitude and phase information of the 'carved-out' portion of the mid-IR field to be sampled.

In her thesis, Christina Hofer studied in detail the nonlinear processes underlying EOS and developed measurement methods to characterize and optimize the efficiency of the above-mentioned steps in EOS. Using long-wavelength, high-power gate pulses, she has demonstrated high-fidelity electric field sampling of octave-spanning mid-infrared waves in the 18–39 THz (600–1300 cm^{-1}) spectral region over an unprecedented dynamic range of amplitudes from the MV/cm level down to a few mV/cm. The field-measurement sensitivity approached within a factor of 4 the ultimate detection limit of capturing all photons in the temporal gate.

The high sensitivity of EOS also enabled a highly sensitive measurement of the fluctuations of the MHz-rate IR pulses, by repeatedly sampling the mid-IR waves at nominally equal timing within the waveform (i.e., with constant delay between test wave and gate pulse). A study of the correspondence between the measured—very small—fluctuations of the EOS signal and the actual optical electric field, revealed temporal fluctuations on the order of 1 attosecond for several thousand pulses, on time scales shorter than 0.1 ms. This provided the basis for a novel EOS-based delay calibration between two oscillators—*electro-optic delay tracking*.

These advances constitute major contributions toward establishing field-resolved IR spectroscopy for real-world biomedical applications.

Garching and Munich,
Germany/Budapest, Hungary
August 2022

Prof. Ferenc Krausz

Abstract

Electro-optic sampling (EOS) is a powerful method for the characterization of electric fields with frequencies in the range of $\approx$ 1–300 THz. For mid-infrared (MIR) radiation (2–20 μm), it can be understood as a two-step process: First, a sub-MIR-cycle visible/infrared gate pulse generates sum and/or difference frequency radiation with the light field under investigation in a nonlinear medium. Second, the newly generated frequencies are detected in a heterodyne scheme with the transmitted gate pulse serving as the local oscillator. Scanning the delay of the gate pulse with respect to the MIR waveform results in a signal proportional to the incident MIR field, with the spectral response depending on the gate-pulse duration and phasematching in the detection crystal. The nonlinear frequency conversion on the one hand transfers the detection to the near-infrared spectral range, affording the use of low-noise photodetectors. On the other hand, it limits the detection efficiency and subjects it to a trade-off against bandwidth.

Our research group develops high-power ultrashort-pulsed laser sources for field-resolved infrared spectroscopy. Explicitly, nonlinearly post-compressed femtosecond lasers are used both to drive the generation of waveform-stable MIR light for molecular-sample excitation as well as for obtaining gate pulses for EOS of the full macroscopic sample response. To maximize the sensitivity of our field-resolved spectrometers, this thesis studied the photon detection efficiency of EOS for MIR radiation with wavelengths in the 6–18-μm range in experiment and theory.

Three different types of gate pulses were investigated experimentally: First, the EOS detection efficiency was characterized for gate pulses with 1030-nm central wavelength, generated by an Yb-thin-disk oscillator. Limited by multi-photon-absorption-caused damage of the GaSe crystal, with an average gate-pulse power of 450 mW, a conversion efficiency of 2% from the MIR into sum-frequency photons was achieved in a 500-μm-thick detection crystal. Accounting for Fresnel reflections at the crystal and losses in the heterodyne detection, up to 0.76% of the incident MIR photons arrived at the balanced diodes. Together with mW-level MIR average powers, this resulted in 13 orders of magnitude frequency-domain intensity dynamic range at 9-μm wavelength for a measurement time of 16s and a scan range of 3.3 ps. However, phase-mismatch limited the –20 dB spectral width to 1.2 μm. Using a

85-μm-thick GaSe crystal, the full MIR spectrum of the source, spanning from 6.6 to 10.7 μm at –20 dB, was detected, while trading in two orders of magnitude in peak dynamic range. In our research group, the prototype field-sensitive spectrometer with this record detection efficiency and dynamic range is currently being used for fingerprinting real-world biomedical samples, with up to 40 times higher molecular detection sensitivity than commercial Fourier-transform infrared spectrometers.

Due to the dispersion of GaSe, the trade-off between detection efficiency and spectral coverage is mitigated for longer-wavelength gate pulses. Using gate pulses centered at 1550 nm wavelength from an Er-fiber laser and a 300-μm-thick crystal, a comparable detection dynamic range and bandwidth as with the 85-μm-thick crystal at 1030-nm was achieved, despite the lower gate-pulse power of 120 mW. This performance enables high-sensitivity spectroscopic measurements, when employing the Er-laser in a dual-oscillator fast-scanning mode ($\approx$ 1 kHz scan rate), avoiding low-frequency noise sources.

In addition to the broader phasematching bandwidth, choosing a longer gate-pulse wavelength also increases the detection-crystal damage threshold due to reduced multi-photon absorption, allowing for the use of higher gate-pulse powers and, consequently, enhancing the nonlinear interaction. This benefit was harnessed in the investigation of limitations to the detection efficiency with 1.9-W gate pulses from a Tm-fiber laser at 1965 nm central wavelength, comparing several EOS crystal thicknesses with respect to detection efficiency and spectral coverage. Traces measured with 100 to 300-μm-thin crystals closely resemble the incident field, spanning from 8.1 μm to 14.2 μm at –10 dB, with a conversion efficiency from the MIR into sum-frequency photons of up to $\approx$ 10%. Using a 500-μm-thick GaSe crystal, more than 20 % of the MIR photons from a 3-μm spectral band around 9.3 μm were upconverted. Further increasing the crystal thickness resulted in saturation of the depletion, explained by temporal walk-off and reduced peak powers due to dispersion. The overall number of detectable MIR photons of $\approx$ 6.4% from within the detection crystal and interaction time window, together with mW-level MIR powers, lead to a peak intensity dynamic range $>10^{14}$. At the same time, the detection bandwidth was twice as broad as for the 1030-nm gate pulses in the efficiency-optimized configuration, thus spanning $\approx$ 5 μm at –20 dB. Despite the MIR depletion upon detection, the EOS signal scaled linearly with the field strength for average photon numbers between $>10^3$ and $>10^{17}$ per second within our measurement accuracy, because enough MIR photons stay available for nonlinear interaction. The multi-percent-level conversion efficiency allows for characterization of waveforms with an average of 22 photons inside the detection crystal in a 2.2-ms-long integration time window per temporal element. The combination of sensitivity, dynamic range and spectral coverage finds application, e.g., in broadband vibrational spectroscopy, where the minimum detectable concentration is only a factor of $\approx$ 4 higher than what would be possible when detecting all incident MIR photons. Furthermore, the detection bandwidth allows for the simultaneous measurement of multiple molecular species with spectrally wide-spread absorption lines.

Employing the high detection dynamic range, a further study in the frame of this thesis concerned the use of EOS as a highly sensitive characterization technique for the stability and reproducibility of the MIR waveform and, therefore, for the control over optical fields. These capabilities were demonstrated by measuring the temporal fluctuations of the EOS trace, resulting in a record-low timing jitter of $<$ 10 as over billions of pulses. A theoretical model simulating the chain of nonlinear processes from the laser frontend to EOS detection confirmed the measured values, identifying intensity noise of the modelocked oscillator front-end as the main source of the remaining MIR waveform instabilities. These jitter values were ≈ 3 orders of magnitude above the field fluctuations expected from a shot-noise-limited driving pulse train.

Acknowledgements

After spending more than 6 years at 'attoworld' and 4.5 years on my Ph.D. thesis, I want to thank everyone who contributed in one way or another and who made this time go by ultrafast and with many memorable insights and moments.

First and foremost I thank Prof. Ferenc Krausz for the opportunity to work in his unique group. It was a pleasure to do research surrounded by very clever and nice people and with seemingly unlimited opportunities to explore the limitations of what is physically possible. A heartfelt 'thank you' to Dr. Ioachim Pupeza for welcoming me in his research group back in the cavity days. Thank you for countering my realism with excitement for exploring the ultimate limits in all our heated yet instructive discussions.

I also thank Prof. Alfred Leitenstorfer for being the second examiner for this thesis and for providing us (together with the respective Ph.D. students) with the highly nonlinear fibers. In addition, I thank Prof. Ilka Brunner and Prof. Monika Aidelsburger for being part of my defense committee. Thanks to the IMPRS-APS graduate school for their (financial) support and to Dr. Nicholas Karpowicz for being my secondary advisor and for your insights to the fundamentals of EOS.

Thank you to Matthias Knorr and Prof. Rupert Huber from Regensburg for the collaboration on suppressing the multiple reflections from GaSe using diamond.

Thank you Dr. Ioachim Pupeza, Dr. Alexander Weigel, Dr. Philipp Sulzer, Dr. Theresa Buberl, Daniel Gerz, Dr. Maximilian Högner, Ann-Kathrin Raab, Lena Grosswindhager and Dr. Stephan Meighen-Berger for taking the time to proof-read (parts of) this thesis.

Of all those who supported me scientifically, I want to point out specifically:

Daniel Gerz, for the wonderful collaboration, collegiality and the discussions during our work at the laser system 'Infrasampler 2.0'. Thank you for your never-ending patience for this laser (and for me); Dr. Thomas Butler, Dr. Julia Gessner and Dr. Jia Xu for their work on that laser system and the joint THz experiments and the collaboration partners from Jena for their tireless support.

The 'Cavity-Team', meaning Dr. Ernst Fill, Dr. Maximilian Högner, Ann-Kathrin Raab, Lukas Fürst and Dr. Philipp Sulzer for their (moral) support and the uplifting coffee breaks during the 'last two weeks' of my work in the lab. Thank you Ernst for sharing your seemingly infinite knowledge, for me especially regarding spectroscopy, Maxi, for all of your input and support regarding coding issues (and for still knowing all the 'old' people). Thank you Philipp for eagerly sharing your EOS expertise toward to end of my Ph.D., thus substantially contributing to my understanding.

Dr. Syed Ali Hussain, Dr. Wolfgang Schweinberger and Dr. Marinus Huber for the good collaboration at Infrasampler 1.2 (and even before that). Thanks Ali for the great insights that we developed on the waveform stability, Wolfgang for your input on any imaginable physical order technical question and Marinus for all insights into noise, the FTIR (and for all the fun social activities).

Philip Jacob and Dr. Alexander Weigel for the good collaboration at Infrasampler 1.5 and Leon Helms for insights we had regarding the THz-EOS during his Master's thesis.

Dr. Vladimir Pervak and his team for providing the custom optics with very ambitious specifications for all of the experiments I was involved in.

Almost finally, some less scientific 'thank you's' go to Dr. Theresa Buberl, who always had an open ear as an office and roommate and with whom I share great memories from extensive Ph.D. hat building sessions and all other social activities. Thanks to the other 'girls', namely Dr. Cristina Leonardo (also for proving that chocolate is the ultimate H-bond), Dr. Gaia Barbiero, Dr. Kathrin Mohler, Dr. Liudmila Voronina, Dr. Sonja Tauchert, Masa Bozic, ... for all the amazing dinners and enjoyable evenings. And thank you to Dr. Stephan Heinrich, Ann-Kathrin Raab and Dr. Markus Pötzlberger for discussions on everything in my mother tongue.

A special thank you goes to everyone working in administration—without them many things would be close to impossible: Dr. Nils Haag and Klaus Franke (also for the music sessions), Kerstin Schmidt-Buchmann, Monika Wild, Katharina Adler, Dr. Martin Gross, Dr. Hans Wirth, Florian Saran, Mr. Öhm and Mr. Krapfl. Thank you to all the 'old', former and new group members, who contributed to a great work environment, making me look forward to go to the lab or office to work, spend time together and who were always supportive. I will cherish all the amazing memories we made together.

Danke an Stephan und alle anderen Freunde, egal ob nah oder fern, die meine mitunter arbeitsbedingten Launen erduldet und mich durch die letzten Jahre begleitet haben.

Zu guter letzt, ein unendlicher Dank an meine Eltern und die ganze Familie, dass ihr mir diesen (sehr) langen Ausbildungsweg (und vieles sonst) ermöglicht habt, mich immer unterstützt und bei aller Abstraktheit der Themen doch nicht den Glauben verloren habt, dass 'des scho was Gscheid's is, was die da lernt'.

Contents

Acronyms

DFG	Difference-frequency generation
EOS	Electro-optic sampling
FROG	Frequency-resolved optical gating
FTIR	Fourier-transform infrared
FWHM	Full-width-at-half-maximum
GDD	Group-delay dispersion
IDT	Interferometric delay tracking
IPDFG	Intra-pulse difference frequency generation
IRF	Instrument response function
LPF	Long-pass filter
MIR	Mid-infrared
NIR	Near-infrared
RIN	Relative intensity noise
SFG	Sum-frequency generation
SNR	Signal-to-noise ratio
SPF	Short-pass filter

Chapter 1
Introduction and Motivation

Mode-locked lasers generate trains of pulses with reproducible intensity envelopes and with an exceptional confinement of the optical energy to durations as short as several femtoseconds. The brevity of these light flashes allowed, e.g., for the study of the motion of atomic nuclei during chemical reactions, establishing the field of femtochemistry [1]. The additional measurement and control of the phase of the optical carrier with respect to the temporal intensity envelope of ultrashort pulses [2–5] brought about the concept of the frequency comb: A train of equidistant, phase-coherent laser pulses corresponds to a comb of spectral lines in the frequency domain, with a frequency offset describing the pulse-to-pulse evolution of the waveform with respect to the envelope and a frequency spacing given by the pulse repetition rate. In this framework, optical oscillations at THz and PHz frequencies serve as frequency standards, enabling unprecedented precision in the measurement of natural constants [6, 7]. Various noise sources in mode-locked lasers [8–10] broaden the comb linewidths and require active or passive stabilization to achieve reproducible waveforms.

In time-domain metrology, the control of the optical electric field enables the study of light-matter interaction on the fundamental level of optical-field-induced dynamic charge separation. Characterizing the electric field on this time scale requires nonlinear optical processes, using variably-delayed, sub-cycle events (such as ultra-short gate pulses), providing the temporal resolution. The generation of waveform-stable laser pulses in the near-infrared (NIR) and visible spectral region enables the study of bound electron dynamics in matter on attosecond timescales [11], using field-resolved methods like attosecond streaking or—very recently—(nonlinear) current sampling [12] and electro-optic sampling [13].

© The Author(s), under exclusive license to Springer Nature Switzerland AG 2022

C. Hofer, *Detection Efficiency and Bandwidth Optimized Electro-Optic Sampling of Mid-Infrared Waves*, Springer Theses,
https://doi.org/10.1007/978-3-031-15328-0_1

Intra-pulse Difference Frequency Generation (IPDFG) and Electro-optic Sampling (EOS)

Combining the generation of waveform-stable pulses in the mid-infrared (MIR) spectral region (2–20 μm) *via Intra-pulse difference frequency generation (IPDFG)* and their field-sensitive pulse characterization using electro-optic-sampling (EOS) [14, 15] allows to probe a wide range of dynamics in solids [16–18] and molecules [19, 20]. The sub-cycle temporal resolution of EOS was recently employed to study squeezing [21] and the vacuum fluctuations of the ground state of electromagnetic radiation [22–24].

In IPDFG, the difference of two frequencies within the same laser pulse is generated via second-order nonlinear mixing, resulting in a waveform that is passively locked in phase with respect to the envelope of the driving pulse [25–27]. The driving pulse envelope thus naturally affords a temporal reference for precision measurements without the need for stabilization of the repetition rate.

EOS makes use of powerful, ultra-short laser pulses and employs second-order perturbative nonlinear processes to sample sub-opticle-cycle portions of the electric field of light. For MIR radiation, it can be most conveniently described as a two-step process [28], combining nonlinear upconversion via sum and difference frequency mixing using a NIR gate pulse, with subsequent heterodyne detection of the newly generated frequencies. As these sum and difference frequency components are polarized perpendicularly to the gate pulse, they change the polarization state in their spectral overlap region [29]. This change is detected with an ellipsometry setup, providing a signal which is proportional to the electric field of the MIR wave.

The spectral response of the detection depends, among other factors, on the duration of the gate pulse and on the phasematching characteristics of the nonlinear medium. On the one hand, the nonlinear frequency conversion transfers signal detection to the NIR spectral region, where low-noise photodetectors are available. On the other hand, the detection efficiency is limited by the efficiency of the nonlinear upconversion process and the measured spectrum is limited in bandwidth due to phasematching in the detection crystal. The research work presented in this thesis was driven by the motivation to explore these trade-offs in electro-optic detection and, thereby, advance the measurement of optical electric fields toward their fundamental limits.

MIR Spectroscopy and EOS: The State of the Art

The gold-standard method used for spectroscopic analysis of various molecular species with absorption lines in the spectral region from 2 to 20 μm is Fourier-transform infrared (FTIR) spectroscopy [30]. This technique employs incoherent, thermal light sources and detectors, or mercury-cadium-telluride (MCT) detectors which require active cooling to achieve high sensitivity [31, 32]. Recently, quantum cascade lasers [33] and MIR sources based on IPDFG, which uniquely combine a broad spectral bandwidth and high brilliance [34–39] while being relatively compact, were developed as significantly more brilliant light sources covering the molecular fingerprint region. The spatial coherence of these sources allows to e.g., increase the interaction length with a sample and thus enhance the spectroscopic sensitivity. FTIR spectroscopy records an interferogram, from which the spectrum and—depending on the measurement geometry—the spectral phase change caused by transmission through a sample is determined. Using two frequency combs, time-domain spectroscopy of the free-induction decay is performed with fast data acquisition and over nanosecond temporal delay ranges, resulting in unprecedented spectral resolution [40, 41], also in the MIR spectral region [42]. Furthermore, upconversion spectroscopy employs the advantage of using low-noise detectors in the NIR [43–45]. This technique uses sum-frequency generation similar to EOS, but typically with narrowband NIR lasers for the nonlinear frequency conversion.

EOS is a well-established technique for field-sensitive measurements in various spectral regions. It was first demonstrated for the characterization of electrical transients more than 30 years ago [46]. In a free-space geometry, it was then applied for measurements of few-THz waves [47] and also time-domain, electric-field-sensitive spectroscopy [48], typically using gate pulses at 800 nm central wavelength from titanium-sapphire oscillators and zinc telluride as the nonlinear crystal. For few-THz waves, EOS is best described using the Pockels-effect, namely a polarization rotation of the gate pulse caused by the quasi-static THz field. More recently, field-resolved characterization of few-cycle laser pulses with ever-shorter wavelengths was made possible by the advancement of femtosecond laser technology [13, 49–52].

While single-shot methods of EOS were demonstrated in the THz spectral region [53, 54], in its standard, multi-shot implementation EOS relies on the pulse-to-pulse reproducibility of the characterized waveform. Thus, MIR radiation generated via intra-pulse difference frequency generation (IPDFG) or optical rectification (OR) serves as an ideal source for field-resolved detection with EOS, suitable e.g., for application in spectroscopy in the molecular fingerprint region [19, 20]. In comparison to time-integrating techniques, such as direct absorption or FTIR spectroscopy, where the excitation pulse impinges on the detector at all delays, temporal gating in EOS allows for filtering the weak molecular response and for adjusting the sensitivity of the detection after the main pulse. Furthermore, detection of a signal proportional to the electric field instead of the MIR intensity results in reduced requirements for the detector dynamic range. This is particularly advantageous when characterizing strongly attenuating samples [55].

In the frame of this thesis, the converison efficiency of MIR into sum-frequency photons in EOS, and hence the number of detectable photons and measurement dynamic range (previously investigated in [56, 57] for THz radiation) were optimized for the first time by means of high gate-pulse powers and in conjunction with thick electro-optic crystals [58]. Using gate pulses at different central wavelengths, the conversion efficiency versus bandwidth trade-off was investigated theoretically and experimentally. Reaching the detection of several percent of the incident MIR photons while maintaining an octave-spanning spectral response, the optimized detection system allows for the spectroscopic analysis of various molecular species with unprecedented field sensitivity.

Furthermore, the measurement sensitivity of EOS makes it suitable for the characterization of the stability of the MIR waveforms. In comparison to well-established methods for the measurement of the pulse-to-pulse evolution of the waveform under the intensity envelope, like e.g., f–2f interferometry [4] or above-threshold ionization [59] for characterizing the electric field on a shot-to-shot basis, the signal scales linearly with the characterized field and does not require additional steps of spectral broadening, which add noise. Contrary to recent investigations on the phase noise of waveforms generated *via* IPDFG with a shot-noise limited source [60, 61], temporal jitter of the pulse envelope is not considered here, because the gate pulse, derived from the same source in the experiments presented here, acts as a temporal reference. Furthermore, temporally chirping the infrared waveform allows for a wavelength-dependent analysis of the waveform stability. In this work, the correspondence between the waveform jitter measured with EOS and the field fluctuations was investigated, tracing the main source of fluctuations back to relative intensity noise of the laser front-end.

Outline of This Thesis

Chapter 2 comprises the theoretical background, namely linear and perturbative nonlinear optical processes employing short laser pulses. An introduction to nonlinear crystals, phasematching and a theoretical description of EOS serve as the basis for the simulation results presented in Chap. 3. These calculations focus on the dependence of phasematching and the spectral response of EOS on the gate pulse parameters and detection crystal thickness. They also investigate differences in the detection efficiency when phasematching sum or difference frequency generation in thick electro-optic media. Furthermore, the significance of the gate pulse phase for the wavelength dependent EOS signal is illustrated.

The experimental results are presented in Chap. 4. First, the detection efficiency of EOS is optimized for 1-μm gate pulses and the trade-off between detection efficiency and bandwidth, as well as the issue of multiple reflections from the detection crystal are discussed. Measuring the MIR power transmitted through the detection crystal as a function of the gate-pulse delay is introduced as a method to characterize the upconversion efficiency. In addition, the application of high-sensitivity EOS for the characterization of the stability of the MIR waveforms generated via IPDFG is presented. The resulting few-as jitter values are traced back to intensity fluctuations of the IPDFG driving laser employing a theoretical model of the involved nonlinear processes. Secondly, longer-wavelength gate pulses are used to overcome the severe efficiency and bandwidth trade-off, allowing for the use of thicker detection materials, which enable the increase of the time-window before the first multiple reflection.

Finally, EOS with percent-level detection efficiency is demonstrated with high-power, 2-μm gate pulses. The significant depletion of the MIR field upon detection necessitates a characterization of the scaling of the signal amplitude with the MIR field strength and the gate pulse power, showing now significant deviation from a linear behavior for mW to fW MIR powers within our measurement accuracy. In linear vibrational spectroscopy, the unprecedented detection efficiency allows for the field-sensitive characterization of broadband molecular emissions only a factor of five stronger than what would be achievable when detecting all infrared photons.

Chapter 5 summarizes the work and gives an outlook on future developments of field-sensitive spectroscopy towards shorter MIR and even longer gate pulse wavelengths. Furthermore, the application of high-dynamic-range EOS for gas-phase absorption spectroscopy is addressed.

References

1. Zewail AH (2000) Femtochemistry: atomic-scale dynamics of the chemical bond. J Phys Chem A 104:5660–5694. ISSN: 1089-5639
2. Reichert J, Holzwarth R, Udem T, Hänsch TW (1999) Measuring the frequency of light with mode-locked lasers. Opt Commun 172:59–68. ISSN: 00304018
3. Telle HR et al (1999) Carrier-envelope offset phase control: a novel concept for absolute optical frequency measurement and ultrashort pulse generation. Appl Phys B 69:327–332. ISSN: 0946-2171
4. Apolonski A et al (2000) Controlling the phase evolution of few-cycle light pulses. Phys Rev Lett 85:740–743
5. Jones DJ et al (2000) Carrier-envelope phase control of femtosecond mode-locked lasers and direct optical frequency synthesis. Science 288:635–640. ISSN: 1095-9203
6. Udem T, Holzwarth R, Hänsch TW (2002) Optical frequency metrology. Nature 416:233–237. ISSN: 1476-4687
7. Hänsch TW (2006) Nobel lecture: passion for precision. Rev Modern Phys 78:1297–1309. ISSN: 0034-6861
8. Paschotta R (2004) Noise of mode-locked lasers (Part I): numerical model. Appl Phys B 79:153–162. ISSN: 0946-2171
9. Paschotta R (2004) Noise of mode-locked lasers (Part II): timing jitter and other fluctuations. Appl Phys B 79:163–173. ISSN: 0946-2171
10. Paschotta R, Schlatter A, Zeller SC, Telle HR, Keller U (2006) Optical phase noise and carrier-envelope offset noise of mode-locked lasers. Appl Phys B 82:265–273. ISSN: 0946-2171
11. Krausz F, Ivanov M (2009) Attosecond physics. Rev Modern Phys 81:163–234. ISSN: 0034-6861
12. Sederberg S et al (2020) Attosecond optoelectronic field measurement in solids. Nat Commun 11:1–8. ISSN: 2041-1723
13. Keiber S et al (2016) Electro-optic sampling of near-infrared waveforms. Nat Photon 10:159–162. ISSN: 1749-4885
14. Leitenstorfer A, Hunsche S, Shah J, Nuss MC, Knox WH (1999) Detectors and sources for ultrabroadband electro-optic sampling: experiment and theory. Appl Phys Lett 74:1516–1518. ISSN: 0003-6951
15. Sell A, Leitenstorfer A, Huber R (2008) Phase-locked generation and field-resolved detection of widely tunable terahertz pulses with amplitudes exceeding 100 MV/cm. Opt Lett 33:2767. ISSN: 0146-9592
16. Huber R et al (2001) How many-particle interactions develop after ultrafast excitation of an electron-hole plasma. Nature 414:286–289. ISSN: 1476-4687
17. Kröll J et al (2007) Phase-resolved measurements of stimulated emission in a laser. Nature 449:698–701. ISSN: 1476-4687
18. Schubert O et al (2014) Sub-cycle control of terahertz high-harmonic generation by dynamical Bloch oscillations. Nature Photon 8:119–123. ISSN: 1749-4885
19. Kowligy AS et al (2019) Infrared electric field sampled frequency comb spectroscopy. Science Adv 5:eaaw8794
20. Pupeza I et al (2020) Field-resolved infrared spectroscopy of biological systems. Nature 577:52–59. ISSN: 1476-4687

21. Riek C et al (2017) Subcycle quantum electrodynamics. Nature 541:376–379
22. Riek C et al (2015) Direct sampling of electric-field vacuum fluctuations. Science 350:420–423. ISSN: 1095-9203
23. Moskalenko AS, Riek C, Seletskiy DV, Burkard G, Leitenstorfer A (2015) Paraxial theory of direct electro-optic sampling of the quantum vacuum. Phys Rev Lett 115. ISSN: 0031-9007
24. Benea-Chelmus I-C, Settembrini FF, Scalari G, Faist J (2019) Electric field correlation measurements on the electromagnetic vacuum state. Nature 568:202–206. ISSN: 1476-4687
25. Baltuška A, Fuji T, Kobayashi T (2002) Controlling the carrier-envelope phase of ultrashort light pulses with optical parametric amplifiers. Phys Rev Lett 88:133901. ISSN: 0031-9007
26. Boyd RW (2008) Nonlinear optics, 3rd edn. Academic Press, Burlington
27. Krauss G et al (2011) All-passive phase locking of a compact Er:fiber laser system. Opt Lett 36:540–542. ISSN: 1539-4794
28. Gallot G, Grischkowsky D (1999) Electro-optic detection of terahertz radiation. J Opt Soc Am B 16:1204. ISSN: 0740-3224
29. Sulzer P et al (2020) Determination of the electric field and its Hilbert transform in femtosecond electro-optic sampling. Phys Rev A 101:033821. ISSN: 1050-2947
30. Griffiths PR, de Haseth JA (2007) In: Griffiths P, De Haseth JA, Winefordner JD (eds) Fourier transform infrared spectrometry, 2nd edn. Wiley-Interscience, Hoboken. ISBN: 978-0-471-19404-0
31. Rogalski A (2003) Infrared detectors: status and trends. Progr Quant Electron 27:59–210. ISSN: 00796727
32. Rogalski A (2012) History of infrared detectors. Opto-Electron Rev 20. ISSN: 1896-3757
33. Villares G, Hugi A, Blaser S, Faist J (2014) Dual-comb spectroscopy based on quantum-cascade-laser frequency combs. Nat Commun 5:5192
34. Pupeza I et al (2015) High-power sub-two-cycle mid-infrared pulses at 100 MHz repetition rate. Nat Photon 9:721–724. ISSN: 1749-4885
35. Zhang J et al (2018) Multi-mW, few-cycle mid-infrared continuum spanning from 500 to 2250 cm-1. Light Sci Appl 7:17180
36. Seidel M et al (2018) Multi-watt, multi-octave, mid-infrared femtosecond source. Sci Adv 4:eaaq1526
37. Gaida C et al (2018) Watt-scale super-octave mid-infrared intrapulse difference frequency generation. Light Sci Appl 7:94
38. Butler TP et al (2019) Watt-scale 50-MHz source of single-cycle waveform-stable pulses in the molecular fingerprint region. Opt Lett 44:1730–1733. ISSN: 1539-4794
39. Wang Q et al (2019) Broadband mid-infrared coverage (2–17 μm) with few-cycle pulses via cascaded parametric processes. Opt Lett 44:2566–2569. ISSN: 1539-4794
40. Coddington I, Swann WC, Newbury NR (2010) Time-domain spectroscopy of molecular free-induction decay in the infrared. Opt Lett 35:1395. ISSN: 0146-9592
41. Coddington I, Newbury N, Swann W (2016) Dual-comb spectroscopy. Optica 3:414. ISSN: 2334-2536
42. Lanin AA, Voronin AA, Fedotov AB, Zheltikov AM (2014) Time-domain spectroscopy in the mid-infrared. Sci Rep 4:6670
43. Johnson TA, Diddams SA (2012) Mid-infrared upconversion spectroscopy based on a Yb:fiber femtosecond laser. Appl Phys B 107:31–39. ISSN: 0946-2171
44. Tidemand-Lichtenberg P, Dam JS, Andersen HV, Høgstedt L, Pedersen C (2016) Mid-infrared upconversion spectroscopy. J Opt Soc Am B 33:D28. ISSN: 0740-3224
45. Lehmann L et al (2019) Single photon MIR upconversion detector at room temperature with a PPLN ridge waveguide. Opt Exp 27:19233–19241. ISSN: 1094-4087
46. Valdmanis J, Mourou G (1986) Subpicosecond electrooptic sampling: principles and applications. IEEE J Quant Electron 22:69–78. ISSN: 0018-9197
47. Wu Q, Zhang X-C (1995) Free-space electro-optic sampling of terahertz beams. Appl Phys Lett 67:3523–3525. ISSN: 0003-6951
48. Lu ZG, Campbell P, Zhang X-C (1997) Free-space electro-optic sampling with a high-repetition-rate regenerative amplified laser. Appl Phys Lett 71:593–595. ISSN: 0003-6951

49. Wu Q, Zhang X-C (1997) Free-space electro-optics sampling of mid-infrared pulses. Appl Phys Lett 71:1285–1286. ISSN: 0003-6951
50. Huber R, Brodschelm A, Tauser F, Leitenstorfer A (2000) Generation and fieldresolved detection of femtosecond electromagnetic pulses tunable up to 41 THz. Appl Phys Lett 76:3191–3193. ISSN: 0003-6951
51. Kübler C, Huber R, Tübel S, Leitenstorfer A (2004) Ultrabroadband detection of multi-terahertz field transients with GaSe electro-optic sensors: approaching the near infrared. Appl Phys Lett 85:3360–3362. ISSN: 0003-6951
52. Mahony TS, Bender DA, Sinclair MB, Brener I (2013) Mid-infrared time-domain spectroscopy system with carrier-envelope phase stabilization. Appl Phys Lett 103:181111. ISSN: 0003-6951
53. Jiang Z, Zhang X-C (1998) Electro-optic measurement of THz field pulses with a chirped optical beam. Appl Phys Lett 72:1945–1947. ISSN: 0003-6951
54. Shan J et al (2000) Single-shot measurement of terahertz electromagnetic pulses by use of electro-optic sampling. Opt Lett 25:426. ISSN: 0146-9592
55. Huber M et al (2020) Optimum sample thickness for trace analyte detection with field-resolved infrared spectroscopy. Anal Chem 92:7508–7514
56. Ibrahim A et al (2016) Ultra-high dynamic range electro-optic sampling for detecting millimeter and sub-millimeter radiation. Sci Rep 6:23107
57. Szwaj C et al (2016) High sensitivity photonic time-stretch electro-optic sampling of terahertz pulses. Rev Sci Instrum 87:103111. ISSN: 1089-7623
58. Kampfrath T, Nötzold J, Wolf M (2007) Sampling of broadband terahertz pulses with thick electro-optic crystals. Appl Phys Lett 90:231113. ISSN: 0003-6951
59. Wittmann T et al (2009) Single-shot carrier-envelope phase measurement of few-cycle laser pulses. Nat Phys 5:357–362. ISSN: 1745-2473
60. Liehl A et al (2019) Deterministic nonlinear transformations of phase noise in quantum-limited frequency combs. Phys Rev Lett 122:203902
61. Liehl A et al (2020) Broadband analysis and self-control of spectral fluctuations in a passively phase-stable Er-doped fiber frequency comb. Phys Rev A 101:023801. ISSN: 1050-2947

Chapter 2
Theoretical Background

2.1 Linear Optics

A single pulse from a mode-locked laser is commonly described mathematically in the frequency domain by an amplitude and phase representation, using the double-sided complex amplitude $\tilde{E}(\omega)$, defined for positive and negative frequencies, with absolute value $A(-\omega) = A(\omega)$ and spectral phase $\varphi(-\omega) = -\varphi(\omega)$

$$\tilde{E}_{\omega} = \tilde{E}(\omega) = A(\omega) \cdot \mathrm{e}^{-\mathrm{i}\varphi(\omega)} \tag{2.1}$$

Its real time-domain electric field can then be calculated as the Fourier transform of this double-sided complex spectral amplitude

$$E(t) = \mathcal{F}\left\{\tilde{E}(\omega)\right\} \tag{2.2}$$

The complex refractive index $\tilde{n}(\omega)$ describes the linear optical effects of dispersion and absorption

$$\tilde{n}(\omega) = n(\omega) - i\kappa(\omega) \tag{2.3}$$

The real part $n(\omega)$ is what is commonly referred to as the refractive index and $\kappa(\omega)$ is the extinction coefficient. Propagation of a laser pulse through a linear medium with that refractive index along the z-direction is then described in the frequency domain as

$$\tilde{E}_{\mathrm{lin}}(\omega) = \tilde{E}_{\mathrm{in}}(\omega)\mathrm{e}^{-i\tilde{k}(\omega)z} = \tilde{E}_{\mathrm{in}}(\omega)\mathrm{e}^{-i\frac{\tilde{n}(\omega)\omega}{c}z} = \tilde{E}_{\mathrm{in}}(\omega)\mathrm{e}^{-i\frac{n(\omega)\omega}{c}z}\mathrm{e}^{-\kappa(\omega)\frac{\omega}{c}z} \tag{2.4}$$

using the definition of the wavenumber $\tilde{k}(\omega) = \frac{\tilde{n}(\omega)\omega}{c}$. The first exponential accounts for linear material dispersion and the second one for attenuation caused by absorption.

© The Author(s), under exclusive license to Springer Nature Switzerland AG 2022
C. Hofer, *Detection Efficiency and Bandwidth Optimized Electro-Optic Sampling of Mid-Infrared Waves*, Springer Theses,
https://doi.org/10.1007/978-3-031-15328-0_2

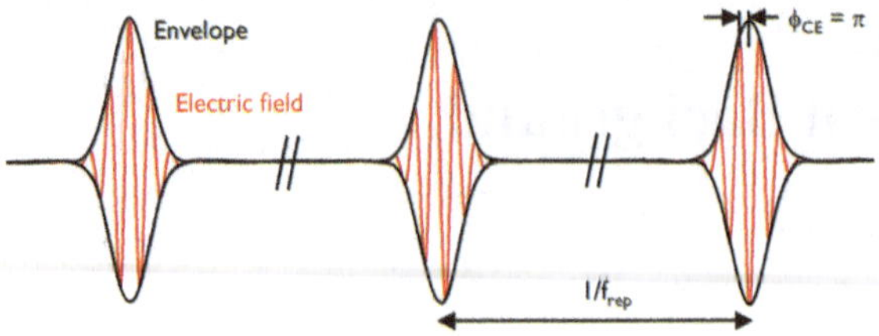

Fig. 2.1 Illustration of the individual pulses of a frequency comb. The temporal envelopes (black) stay constant and are separated in time by the inverse of the repetition frequency f_{rep}. The maximum of the oscillating electric field (red) is offset with respect to the envelope maximum by ϕ_{CE}, which changes from pulse to pulse

The linear spectral phase in the first exponent, $\varphi(\omega) = \frac{n(\omega)\omega}{c} z$, can be Taylor expanded around the central frequency ω_{c} of the laser pulse

$$\varphi(\omega) \approx \varphi_0 + \text{GD} \cdot (\omega - \omega_c) + \frac{\text{GDD} \cdot (\omega - \omega_c)^2}{2} + \cdots \tag{2.5}$$

This allows for the separate description of the individual phase contributions: φ_0 is a constant phase offset for all frequencies, causing a temporal shift of the peak of the oscillating waveform with respect to the envelope (see Fig. 2.1). The group delay (GD) shifts the envelope in time, the group-delay dispersion (GDD) is the lowest-order phase term that causes distortions of the pulse shape.

Frequency Comb Picture and Waveform Stability

For modelocked lasers with MHz to GHz repetition rates, no individual pulses, but the equidistantly spaced pulse-train is considered, as illustrated in Fig. 2.1.

Assuming identical temporal envelopes for each of the pulses, the time-domain electric field E_{fc} is given as [1]

$$E_{\text{fc}}(t) = \text{Re}\left\{\sum_n A_n \text{e}^{-i(\omega_c + n2\pi f_r)t}\right\} \tag{2.6}$$

with the n-th temporal envelope A_n, the central frequency ω_c and the repetition frequency f_r. In the frequency domain, this pulse train corresponds to a series of comb lines, where the n-th mode is calculated as [2]

$$\nu_{\text{n}} = n f_{\text{r}} + f_0 \tag{2.7}$$

The repetition frequency f_{r} is the inverse of the time difference T_{r} between two subsequent laser pulses: $f_{\text{r}} = \frac{1}{T_{\text{r}}}$ and f_0 is the comb offset, determining the pulse to pulse phase slip $\Delta\phi_{\text{ce}}$ of the waveform under the temporal envelope:

$$f_0 = \frac{1}{2\pi} f_{\text{r}} \Delta\phi_{\text{ce}} \tag{2.8}$$

The relative phase ϕ_{ce} between the waveform maximum and the peak of the envelope corresponds to the constant offset φ_0 of the spectral phase in Eq. (2.5). For an offset frequency $f_0 = 0$, the electric field does not shift under the envelope, corresponding to identical waveforms for each laser shot. This reproducibility is one crucial prerequisite for the measurement of the electric field with multi-shot techniques, like the implementation of EOS used in this thesis.

Polarization and Jones-Matrices

The electric field of polarized light can be described as a linear combination of two orthogonal polarization components, using the Jones formalism [3]. Usually, it is split into horizontally (along the x-axis, p-) and vertically (along the y-axis, s-) polarized light. The electric field vector $\tilde{\mathbf{E}}$, with components $\tilde{E}_x$ and $\tilde{E}_y$ is then written as

$$\tilde{\mathbf{E}} = \begin{pmatrix} \tilde{E}_\mathrm{x} \\ \tilde{E}_\mathrm{y} \end{pmatrix} = \begin{pmatrix} A_{0,\mathrm{x}}\, \mathrm{e}^{-i\varphi_\mathrm{x}} \\ A_{0,\mathrm{y}}\, \mathrm{e}^{-i\varphi_\mathrm{y}} \end{pmatrix} \tag{2.9}$$

where $A_{0,\mathrm{x/y}}$ are the spectral amplitudes and $\varphi_{\mathrm{x/y}}$ the spectral phases (see Eq. (2.1)) for each polarization component. For the understanding and theoretical implementation of the ellipsometry setup that is the basis of EOS detection described in Sect. 2.3, the effect of a quarter and a half wave plate on the different polarization components is required. The effect can be calculated using the corresponding *Jones-matrices*. For a quarter wave plate with its fast axis rotated to an angle θ with respect to the horizontal axis (x-axis, or p-polarization axis), the matrix reads

$$M_{\lambda/4} = \mathrm{e}^{-\frac{i\pi}{4}} \begin{pmatrix} \cos^2\theta + i\sin^2\theta & (1-i)\sin\theta\cos\theta \\ (1-i)\sin\theta\cos\theta & \sin^2\theta + i\cos^2\theta \end{pmatrix} \tag{2.10}$$

Similarly, for a half wave plate with its fast axis rotated to an angle θ with respect to the horizontal axis, the matrix reads

$$M_{\lambda/2} = \mathrm{e}^{-\frac{i\pi}{2}} \begin{pmatrix} \cos^2\theta - \sin^2\theta & 2\sin\theta\cos\theta \\ 2\sin\theta\cos\theta & \sin^2\theta - \cos^2\theta \end{pmatrix} \tag{2.11}$$

This allows for the calculation of the phase changes of the individual polarization components *via* multiplication of the respective matrix with the electric field vector.

2.2 Perturbative Nonlinear Optics

Characterizing MIR radiation with linear techniques like (dispersive) FTIR spectroscopy results in its spectral amplitude and the linear phase shift caused by a sample. Gaining information on the temporal evolution of the optical electric field requires nonlinear optical gating processes.

Nonlinear optical effects become observable when the strength of the electric field causes a distortion of the initially harmonic motion of the charges which is caused by light-matter interaction [4]. In this case, the response of a material on the transmitted laser pulse is not sufficiently described by the linear propagation with the complex refractive index in Eq. (2.4), but the interaction depends on higher powers of the electric field. For the effects considered here, the polarization induced by an electric field $E(t)$ can be treated in the frame of perturbation theory, where the higher-order terms in the Taylor series decrease in strength:

$$\begin{aligned} P(t) &= \epsilon_0 \left[\chi^{(1)} E(t) + \chi^{(2)} E(t)^2 + \chi^{(3)} E(t)^3 + \cdots \right] \\ &\equiv P^{(1)}(t) + P^{(2)}(t) + P^{(3)}(t) + \cdots \\ &\equiv P^{(1)} + P^{\mathrm{NL}} \end{aligned} \tag{2.12}$$

This formalism treats the nonlinear terms as perturbations to the linear response and is therefore commonly split into linear and nonlinear contributions [4]. The nth order optical susceptibility $\chi^{(\mathrm{n})}$ indicates the strength of the respective nonlinear effect, where the possible nonlinearities are determined by the crystal symmetry, i.e., second-order effects can only occur in materials without inversion symmetry [4]. In general, it is a tensor describing the interaction between the different input polarisation components. The non-zero entries of this tensor depend on the material symmetry. For a given crystal—and input field geometry, the effective interaction strength of an *instantaneous* nonlinearity can be calculated, reducing the tensor to a scalar quantity [5].

Propagation of a laser pulse's frequency component ω through a nonlinear medium along the z-direction can be described by the first order propagation equation, as derived in [6]:

$$\frac{\partial \tilde{E}_\omega(z)}{\partial z} = -ik_\omega \tilde{E}_\omega(z) - \frac{i\omega}{2n(\omega)\epsilon_0 c} \tilde{P}_\omega^{\mathrm{NL}}(z) \tag{2.13}$$

It assumes small changes of the induced polarization in transverse dimensions on the scale of the central wavelength as it is based on the scalar wave equation. Furthermore, due to polarization-induced changes that are small in comparison to the electric field on this scale, the slowly-evolving wave approximation can be applied [7]. The first term on the right-hand side accounts for linear dispersion and absorption, equivalent to Eq. (2.4). The second term describes nonlinear effects, with the nonlinear polarization in the frequency domain calculated as the Fourier transform of the time-domain sum of the relevant terms:

$$\tilde{P}^{\mathrm{NL}}(\omega) = \mathcal{F}\left\{P^{(2)}(t) + P^{(3)}(t) + \cdots \right\} \tag{2.14}$$

In this work, instantaneous second and third order effects will be used. The second order nonlinear effects of sum-frequency generation (SFG) and difference-frequency generation (DFG) will be discussed in detail in the theoretical description of EOS in the next section. Furthermore, the third-order effect of self-phase modulation (SPM) or temporal optical Kerr-effect will be simulated in Sect. 4.5 as a cycle-averaged-intensity dependent change of the refractive index [4].

2.3 Theoretical Description of EOS

For the detection of few-THz radiation, EOS is commonly described with the Pockels-effect [8]. The THz field causes birefringence or a change of birefringence of a second-order nonlinear material [9], which rotates the polarization of a probe/gate pulse, propagating through the medium at a varied time delay. However, this description assumes that all gate pulse wavelengths experience polarization rotation. This is only the case for THz frequencies much smaller than the spectral width of the gate pulse, where the second-order nonlinearity underlying the Pockels effect does not cause a significant frequency shift [10]. For the detection of MIR or NIR radiation, with frequencies on the order of the spectral width of the gate pulse, the frequency shift is considerable and the sum- and difference-frequency generation picture [11] is more suitable to describe the generation of the electro-optic signal. In this case, polarization changes only occur in the spectral overlap region of the gate pulse and the newly generated frequencies (for a visualization of the polarization rotation see [12, 13]).

Here, a detailed theoretical description of EOS following the sum- and difference-frequency generation picture will be given. For the detection of low-frequency THz radiation, the Pockels-effect picture will be shortly summarized.

2.3.1 Sum- and Difference-Frequency Generation

The two-step model for the description of EOS was first described in [11] and also detailed in [13]. A schematic of a typical detection setup is sketched in Fig. 2.2a.

First, the infrared field to be characterized is converted to higher frequencies, using a short NIR 'gate' or 'probe' pulse (in the following: 'gate'). Specifically, the two fields interact in a second-order nonlinear medium to generate the sum- and difference-frequency fields. In the second step, the newly generated frequencies are detected in a heterodyne scheme with the gate pulse ('local oscillator') in their spectral overlap region (see the sketch of the interacting spectra in Fig. 2.2b), employing an ellipsometry setup consisting of a quarter or half-wave plate and a Wollaston prism, followed by balanced photodiodes.

The input to the nonlinear crystal consists of the superposition of the MIR and the gate fields, each written here in the amplitude and phase representation in the frequency domain (see Eq. (2.1)):

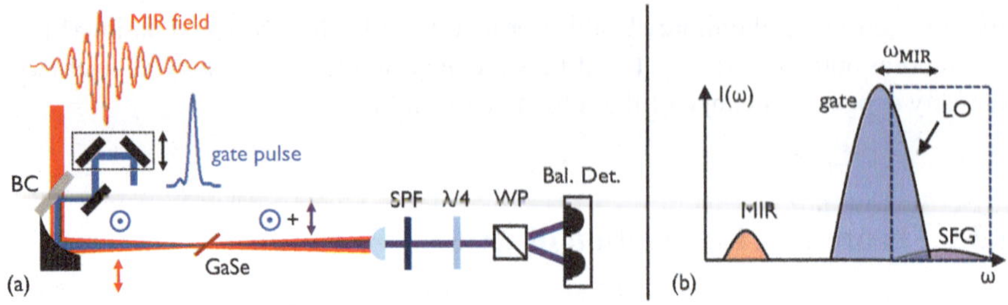

Fig. 2.2 Schematic of a typical EOS setup and sketched spectra of the interacting fields. **a** Typical schematic beampath. The MIR field and variably delayed gate pulse are spatially overlapped using the beam combiner (BC). Arrows and circles indicate the polarization states of the incident fields. Both beams are focused into a nonlinear crystal, i.e., GaSe with an off-axis parabolic mirror. The generated sum-frequency components and the transmitted gate beam are recollimated with a lens and transmitted through a short-pass filter (SPF) for SNR improvement. The quarter-wave plate and Wollaston prism (WP) serve as the ellipsometry setup before balanced heterodyne detection (Bal. Det.), with part of the gate spectrum serving as the local oscillator (LO). **b** Sketch of the frequency-domain spectra involved in the EOS detection. SFG: Sum-frequency generation. The dashed line indicates the transmission through the SPF used for SNR improvement. Gate and sum-frequency spectrum are approximately offset by the central frequency ω_{MIR} of the MIR spectrum

$$\tilde{E}_{MIR}(\omega) = A_{MIR}(\omega) \cdot e^{-i\varphi_{MIR}(\omega)}, \quad \tilde{E}_{gate}(\omega) = A_{gate}(\omega) \cdot e^{-i\varphi_{gate}(\omega)} \tag{2.15}$$

Sum- and difference-frequency mixing is a second-order nonlinear process. Starting from the Taylor expansion for the nonlinear polarization (see Eq. (2.12)) the relevant polarization can be written in a simplified, scalar form for a known crystal geometry and input fields [4, 5]. The nonlinear polarization in the time domain is then given by:

$$P^{(2)}_{EOS}(t) = \epsilon_0 \left[\chi^{(2)} E_{MIR}(t) E_{gate}(t)\right] \tag{2.16}$$

A more detailed description takes into account the tensor nature of the nonlinear susceptibility $\hat{\chi}^{(2)}$ and therefore the varying nonlinearity for different polarization components of the interacting laser fields. This description will be detailed here for GaSe, as this crystal is used in all experiments and in the simulation results presented in Chap. 3. For a second-order nonlinear process, the interaction strength of the field components E'_j and E'_k, that result in a polarization P'_i, for the principal crystal axes i, j, and k, is described by the nonlinear tensor d_{ijk}. This tensor is related to the nonlinear susceptibility according to

$$d_{ijk} = \frac{1}{2}\chi^{(2)}_{ijk} \tag{2.17}$$

In general, this tensor is frequency dependent. However, for frequency ranges away from optical or phonon resonances in the nonlinear material, assuming it to be constant with frequency is a valid assumption [14]. Under this assumption, Kleinman's symmetry for permutation of the mixing field components is valid [4], which allows to write the nonlinear polarization compoenents for sum-frequency mixing as [15]:

$$P_i'(t) = \epsilon_0 2d_{ijk}E_j'(t)E_k'(t) \tag{2.18}$$

The nonlinear tensor can then be written as

$$d_{il} = \begin{bmatrix} d_{11} & d_{12} & d_{13} & d_{14} & d_{15} & d_{16} \\ d_{21} & d_{22} & d_{23} & d_{24} & d_{25} & d_{26} \\ d_{31} & d_{32} & d_{33} & d_{34} & d_{35} & d_{36} \end{bmatrix} \tag{2.19}$$

This allows to write the nonlinear polarization vector for the three principal crystal axes as a function of the field components along those same axes as the matrix multiplication:

$$P_{\mathrm{NL}}' = \begin{bmatrix} P_1' \\ P_2' \\ P_3' \end{bmatrix} = 2\epsilon_0 \cdot d_{il} \cdot \begin{bmatrix} E_1'E_1' \\ E_2'E_2' \\ E_3'E_3' \\ E_2'E_3' + E_3'E_2' \\ E_1'E_3' + E_3'E_1' \\ E_1'E_2' + E_2'E_1' \end{bmatrix} \tag{2.20}$$

Crystal symmetries limit the number of non-zero and independent elements of d_{il}. In the case of GaSe, with its hexagonal structure and crystal class $\bar{6}\,\mathrm{m}^2$, these are: $d_{21} = d_{16} = -d_{22} = -54\,\mathrm{pm/V}$ [16, 17].

As d_{il} is given in the reference frame of the crystal, but the response is calculated for the input field polarization components, the polarization vector has to be transferred back to the laboratory reference frame [15]:

$$P_{\mathrm{NL}}(E) = U\,P_{\mathrm{NL}}'\left(U^T E\right) \tag{2.21}$$

with the transformation matrix U. For a uniaxial crystal such as GaSe, this matrix depends on the two rotation angles θ and φ according to:

$$U = \begin{pmatrix} \cos\varphi\cos\theta & \sin\varphi\cos\theta & -\sin\theta \\ -\sin\varphi & \cos\varphi & 0 \\ \sin\theta\cos\varphi & \sin\varphi\sin\theta & \cos\theta \end{pmatrix} \tag{2.22}$$

The rotation angles are visualized in the sketch of the GaSe crystal in Fig. 2.3a.

With this transformation, the nonlinear polarization for the two orthogonal field polarization directions E_x (p-polarization) and E_y (s-polarization) is [18]:

$$P_x = P_e = 2\epsilon_0\left[2d_{22}\cos^2\theta\cos 3\phi E_x E_y + d_{22}\cos\theta\sin 3\phi E_y^2\right] \tag{2.23}$$

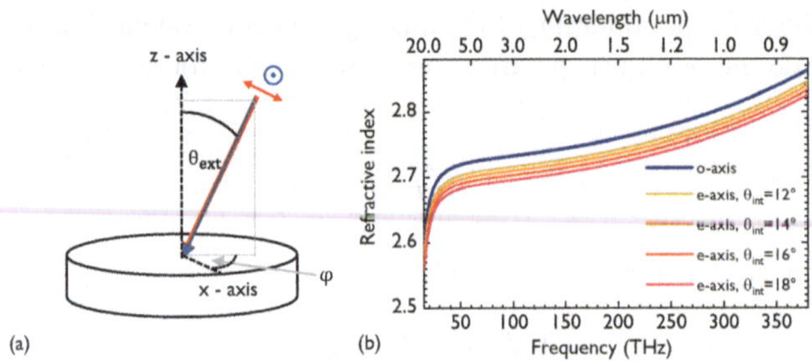

Fig. 2.3 **a** Sketch of the crystal geometry for gallium selenide (GaSe). θ_{ext} is the angle between the optical axis along the z-direction of the nonlinear crystal and the propagation axis of the input beam(s). φ is the angle between the plane spanned by the incoming beams and the z/optical-axis to the crystal x-axis. The arrow and circle indicate the polarization directions of the input beams. **b** Ordinary and extra-ordinary refractive indices in GaSe, the latter for varying internal θ-angles

$$P_y = P_o = 2\epsilon_0 \left[d_{22} \cos^2\theta \cos 3\phi E_x^2 + 2d_{22} \cos\theta \sin 3\phi E_x E_y\right] \tag{2.24}$$

Inserting these nonlinear polarization components into the first-order propagation equation, leads to coupled differential equations for the two polarization axes. The evolution of the s-polarized field component, that propagates on the ordinary axis when assuming a vertical rotation axis for θ, is described in the frequency domain by Conforti et al. [15]:

$$\frac{\partial \tilde{E}_o}{\partial z} = -ik_o\tilde{E}_o - \frac{i\omega}{n_o(\omega)c}\mathcal{F}\left[d_1 E_e^2 + 2d_2 E_e E_o\right] \tag{2.25}$$

and the evolution of the p-polarized field on the extra-ordinary axis evolves according to:

$$\frac{\partial \tilde{E}_e}{\partial z} = -ik_e\tilde{E}_e - \frac{i\omega}{n_e(\omega)c}\mathcal{F}\left[2d_1 E_o E_e + d_2 E_o^2\right] \tag{2.26}$$

The effective nonlinear coefficients follow from the transformation matrix, and are [5]:

$$d_1 = d_{22} \cos^2\theta \cos 3\phi, d_2 = d_{22} \cos\theta \sin 3\phi \tag{2.27}$$

In our experiments, we chose the phasematching configuration for SFG in GaSe, where the gate pulse on the ordinary axis and the MIR field on the extraordinary axis generate the sum-frequency field on the extraordainary axis ('type II phasematching'). Gate and sum-frequency field have to have perpendicular polarizations for the electro-optic signal to be field-dependent when measured with the standard balanced setup. The initial conditions for the nonlinear interaction are therefore

$$E_{o,0} = E_{\text{gate}}, E_{e,0} = E_{\text{MIR}} \tag{2.28}$$

The relevant component of Eq. (2.26) is therefore the mixing term of E_o and E_e, that is maximized for $\varphi = 0°$.

Analogously, phasematching DFG is most efficient for the gate pulse propagating on the ordinary axis and the MIR field on the extraordinary one. The DFG then also has to be generated on the ordinary axis in order to be detectable. Therefore, the mixing term in Eq. (2.25) has to be maximized, which is the case for $\varphi = 30°$. The field evolution throughout the nonlinear crystal according to these propagation equations is implemented in the EOS simulation code used for calculations throughout this thesis, with the main results presented in Chap. 3. Dispersion and phasematching are considered by the linear term of the propagation equation. Cascaded effects of second order at all wavelengths are also included, because the interaction between the different polarization components is calculated after each propagation step.

The second step of the EOS detection is the heterodyne detection of the generated sum- and difference-frequency components with the gate pulse constituting the 'local oscillator'. For a simple description of the scaling laws, only the mixing of three single frequency components, is written here. The sum- or difference-frequency field is therefore [11, 13]:

$$\tilde{E}_{\text{SFG/DFG}}(\omega \pm \Omega) = A_{\text{SFG/DFG}}(\omega \pm \Omega)\mathrm{e}^{-i\varphi_{\text{SFG/DFG}}(\omega\pm\Omega)} \propto i\tilde{E}_{\text{gate}}(\omega) \cdot \tilde{E}_{\text{MIR}}(\Omega) \tag{2.29}$$

Using the amplitude—and phase representation of the input fields, this yields:

$$A_{\text{SFG/DFG}} \propto A_{\text{gate}} \cdot A_{\text{MIR}} \tag{2.30}$$

$$\varphi_{\text{SFG/DFG}} \propto \frac{\pi}{2} + \varphi_{\text{gate}} \pm \varphi_{\text{MIR}} \mp \Omega\tau \tag{2.31}$$

for the relative delay τ between the MIR and gate pulses. Because we typically use >50-μm-thick detection crystals, strongly favouring phasematching of SFG and thus suppressing the DFG components, only the sum-frequency contribution is considered in the following. The simulation results describe the more complicated mixing of all wavelengths and also the full nonlinear response.

In the ideal picture of gate and MIR being perfectly polarized and not experiencing any birefringence of the nonlinear crystal, the generated sum-frequency radiation cannot interfere with the local oscillator field $\mathbf{E}_{\text{LO}}$, as they propagate on orthogonal polarization axes ($\mathbf{e}_x = \mathbf{e}_e$, p-polarization and $\mathbf{e}_y = \mathbf{e}_o$, s-polarization):

$$\tilde{\mathbf{E}}_{\text{LO}} = \begin{pmatrix} 0 \\ \tilde{E}_{\text{LO}} \end{pmatrix}, \tilde{\mathbf{E}}_{\text{SFG}} = \begin{pmatrix} \tilde{E}_{\text{SFG}} \\ 0 \end{pmatrix} \tag{2.32}$$

The superposition of the two fields after the nonlinear crystal, $\tilde{\mathbf{E}}_{\text{het}}$ is therefore given by:

$$\tilde{\mathbf{E}}_{\text{het}} = \begin{pmatrix} \tilde{E}_{\text{SFG}} \\ \tilde{E}_{\text{LO}} \end{pmatrix} \tag{2.33}$$

To allow for interaction of the two fields, they have to be projected onto common polarization axes. This is achieved by using a quarter—or half waveplate [13] and a Wollaston prism ('ellipsometry setup'). The waveplates change the polarization state according to the Jones matrices given in Eqs. (2.10) and (2.11). We typically use the quarter-wave plate configuration, with $\theta_{\lambda/4} = 45°$. For compensation of the chromaticity of the quarter-wave plate and crystal birefringence, we use an additional half-wave plate rotated to a small angle, $\theta_{\lambda/2} \approx 0$. For simplicity, the effect of the second waveplate is not considered for the mathematical description here. After the quarter-wave plate, the electric field vector of the heterodyne signal is therefore:

$$\tilde{\mathbf{E}}_{\text{het,mix}} = \frac{e^{-i\pi/4}}{2}\begin{pmatrix} 1+i & 1-i \\ 1-i & 1+i \end{pmatrix} \cdot \begin{pmatrix} \tilde{E}_{\text{SFG}} \\ \tilde{E}_{\text{LO}} \end{pmatrix} = \frac{1}{2}\begin{pmatrix} \tilde{E}_{\text{SFG}} - i\tilde{E}_{\text{LO}} \\ -i\tilde{E}_{\text{SFG}} + \tilde{E}_{\text{LO}} \end{pmatrix} \tag{2.34}$$

A Wollaston prism then splits this field into the x and y polarization components, that are measured with balanced photodiodes. For $\tilde{E}_{\text{SFG}} = 0$ the total field is the circularly polarized local oscillator field, with equal power to both diodes. For $\tilde{E}_{\text{SFG}} \neq 0$ and at $\tau = 0$, the intensity of each polarization component is:

$$\begin{aligned} I_x &= |A_{\text{SFG}}|^2 + |A_{\text{LO}}|^2 + A_{\text{LO}}A_{\text{SFG}}(e^{i(\varphi_{\text{LO}}-\varphi_{\text{SFG}}-\pi/2)} + e^{-i(\varphi_{\text{LO}}-\varphi_{\text{SFG}}+\pi/2)}) \\ I_y &= |A_{\text{SFG}}|^2 + |A_{\text{LO}}|^2 - A_{\text{LO}}A_{\text{SFG}}(e^{i(\varphi_{\text{LO}}-\varphi_{\text{SFG}}-\pi/2)} + e^{-i(\varphi_{\text{LO}}-\varphi_{\text{SFG}}-\pi/2)}) \end{aligned} \tag{2.35}$$

Subtracting the two signals gives the measured differential intensity:

$$\begin{aligned} \Delta I = I_x - I_y &= 4A_{\text{LO}}A_{\text{SFG}}\sin(\varphi_{\text{LO}} - \varphi_{\text{SFG}}) \\ &= 4A_{\text{LO}}A_{\text{gate}}A_{\text{MIR}}\sin\left(\varphi_{\text{LO}} - \varphi_{\text{gate}} - \varphi_{\text{MIR}} - \frac{\pi}{2}\right) \\ &\approx 4A_{\text{gate}}^2 A_{\text{MIR}}\cos(\varphi_{\text{MIR}}) \end{aligned} \tag{2.36}$$

The last simplification step can only be made for constant amplitude and phase of the gate field at both frequencies ω and $\omega + \Omega$. Furthermore, only a single sum-frequency was considered. In reality, the intensity measurement integrates over all frequencies and potential phase differences reduce the signal strength. These effects are discussed in [13] and will be illustrated in Sect. 3.1. The scaling of the signal strength with the intensity of the gate pulse and the strength of the investigated field is crucial for the experiments described below.

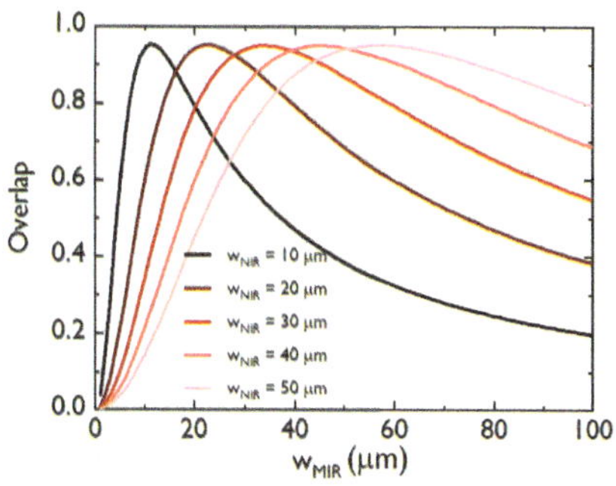

Fig. 2.4 Calculated spatial overlap in EOS calculated as the product of the overlap for SFG and for heterodyne detection. w_{NIR} and w_{MIR} are the $\frac{1}{e^2}$ beam radii of the gate and MIR beams, respectively

Spatial Overlap in EOS

No spatial effects were considered in the theoretical description so far. However, the EOS signal depends on the strength of two interactions: In the first step, SFG depends on the product of the MIR and gate fields, and in the second step, heterodyning requires interaction of the generated sum-frequency field and the local oscillator. For maximum signal strength, the spatial overlap of the interacting fields is therefore crucial. Adapting the formalism presented in [19], the complete overlap integral T_{overlap} is calculated here as the product of the two overlaps, $T_{\mathrm{MIR,NIR}}$ for the sum-frequency generation and $T_{\mathrm{SFG,NIR}}$ for the heterodyning:

$$T_{\mathrm{overlap}} = T_{\mathrm{MIR,NIR}} \cdot T_{\mathrm{SFG,NIR}} = \frac{4 w_{\mathrm{NIR}} w_{\mathrm{MIR}}^2}{\sqrt{w_{\mathrm{NIR}}^2 + w_{\mathrm{MIR}}^2 (w_{\mathrm{NIR}}^2 + 2 w_{\mathrm{MIR}}^2)}} \tag{2.37}$$

The spots of the gate and MIR beams are assumed to be concentric inside the EOS crystal, with $\frac{1}{e^2}$-spot sizes w_{gate} and w_{MIR}, respectively. Using these spot sizes as the only free parameters, the SFG beam is calculated as the product of the interacting beams. The overlap calculated using Eq. (2.37) is shown in Fig. 2.4 for several NIR-beam radii as a function of the MIR beam radius.

The overlap is maximized for $w_{\mathrm{MIR}} \approx w_{\mathrm{NIR}}$, with a slight increase of the optimum ratio to values > 1 for larger w_{NIR}. In addition, the exact MIR beam size is less crucial for larger NIR foci. Typically, due to the focusing properties of Gaussian beams, $w_{\mathrm{MIR}} < w_{\mathrm{gate}}$ applies. In the EOS optimization described below, telescopes are used to optimize this overlap integral. As Eq. (2.37) assumes concentric beams and collinearity, the reduction of the spatial overlap due to spatial walk-off inside the detection crystal is not considered. Equation (2.37) was used to determine the optimum beam sizes in the experimental setups, but is not included in the theoretical model.

2.3.2 Pockels-Effect

The sum- and difference-frequency generation formalism can also be applied for the detection of radiation with few-THz frequencies. In this limit, $\omega_{\text{THz}} \approx 0$, and the generated frequency components are barely shifted with respect to the gate pulse spectrum: $\omega_{\text{SFG}} = \omega_{\text{gate}} \pm \omega_{\text{MIR}} \approx \omega_{\text{gate}}$. However, they are still generated on the perpendicular polarization axis, thus resulting in polarization rotation over (almost) the full gate pulse spectrum.

This polarization rotation is commonly described by the linear electro-optic effect or Pockels effect, where a static electric field changes the principle refractive indices of the nonlinear material and therefore induces or changes the birefringence. A derivation of the electro-optic effect for zincblende materials (like ZnTe and GaP) is given in [9] and the THz EOS signal is calculated in [20]. A discussion of the dependence of the signal on the THz polarization is presented in [21], and polarization sensitive electro-optic detection is discussed in [22, 23].

The mathematical description of the Pockels effect starts from a change of the principle refractive indices by the applied electric field according to [4]:

$$\Delta\left(\frac{1}{n^2}\right)_i = \sum_j r_{ij} E_j \tag{2.38}$$

In analogy to the tensor of the nonlinear coefficients used in the sum-frequency picture, the tensor of the electro-optic coefficients, r_{ij}, quantifies the strength of the nonlinear interaction, in this case the change of the refractive index. Again, crystal symmetries limit the number of independent and non-zero tensor elements. Here, the EOS signal is given for zinc-blende structured crystals, following the calculations for (110)-cut ZnTe in [20]. The same description can be used for (110)-cut gallium phosphide which was used for the THz experiments described in [24]. The only non-zero electro-optic coefficient in GaP is $r_{41} = 0.97\,\text{pm/V}$ [25]. A gate pulse, initially polarized parallel to the (001) or $(\bar{1}10)$ axis of the crystal, experiences a phase retardation, as the crystal acts as a waveplate rotated to 45° with respect to the gate pulse polarization [26]. This phase retardation Γ is caused by the refractive index difference along the rotated principal axes of the nonlinear material of length L, n'_x and n'_y [4]:

$$\Gamma = (n'_x - n'_y)\frac{\omega L}{c} \tag{2.39}$$

In [21], the refractive index changes are given as a function of the THz and gate beam polarization. They showed that the maximum polarization rotation is achieved for both the THz and gate being polarized at 90° to the (001)-axis of the crystal. This leads to [26]:

$$\Gamma_{\text{THz}} = \frac{2\pi L}{\lambda_{\text{gate}}} n^3 r_{41} E_{\text{THz}} \tag{2.40}$$

with the central wavelength λ_{gate} of the gate pulse and the corresponding refractive index n_{gate}. For weak THz field strengths and balancing with a quarter-wave plate rotated to 45°, the measured difference output ΔI of a balanced diode, normalized to the total signal I_{tot} is given by:

$$\frac{\Delta I}{I_{\mathrm{tot}}} \propto \sin(\Gamma_{\mathrm{THz}}) \approx \Gamma_{\mathrm{THz}} \tag{2.41}$$

thus showing the linear correspondence between the THz field and the electro-optic signal.

2.3.3 Comparison to Time-Integrating MIR Characterization Techniques

FTIR spectroscopy is the standard technique for absorption measurements in the MIR spectral region. The scaling laws for the signal strengths and fundamental differences between this time-integrating characterization technique and time-gated EOS are summarized here. The interferometric setup for FTIR spectroscopy is shown in Fig. 2.5a.

Light from a broadband source is split into two paths with a beam splitter (BS) and transmitted through a sample. For recording a spectrogram, one of the paths is variably temporally delayed before the two beams are recombined using a second BS. The delay dependent signal $S_{\mathrm{FTIR}}(\tau)$ is calculated as

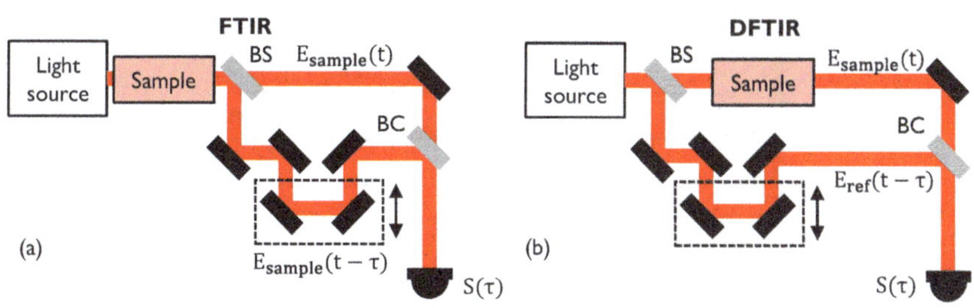

Fig. 2.5 **a** Interferometer setup for FTIR spectroscopy. Light from a broadband source is transmitted through a sample before it is split into two paths with a beam splitter (BS). One of the arms is variably temporally delayed, allowing for the recording of a spectrogram. After recombination at the beam combiner (BC) the intensity of the interfering beams is detected. **b** Interferometer setup for dispersive FTIR, where only one of the beams is transmitted through a sample, thus allowing for the determination of the phase of the sample response

$$
\begin{aligned}
S_{\mathrm{FTIR}}(\tau) &= \int |E_{\mathrm{sample}}(t) + E_{\mathrm{sample}}(t-\tau)|^2 \mathrm{d}t \\
&\propto \int I_{\mathrm{sample}}(t) + I_{\mathrm{sample}}(t-\tau) + E_{\mathrm{sample}}(t) \cdot E_{\mathrm{sample}}(t-\tau) \mathrm{d}t
\end{aligned}
\tag{2.42}
$$

In the dispersive setup [27]—see the schematic in Fig. 2.5b—only one of the beams is transmitted through the sample, while the second one serves as a reference beam, E_{ref}. The resulting signal is given as

$$
\begin{aligned}
S_{\mathrm{DFTIR}}(\tau) &= \int |E_{\mathrm{sample}}(t) + E_{\mathrm{ref}}(t-\tau)|^2 \mathrm{d}t \\
&\propto \int I_{\mathrm{sample}}(t) + I_{\mathrm{ref}}(t-\tau) + E_{\mathrm{ref}}(t-\tau) \cdot E_{\mathrm{sample}}(t) \mathrm{d}t
\end{aligned}
\tag{2.43}
$$

The sample field results from convolution of the reference with the sample response in the time domain: $E_{\mathrm{sample}}(t) = H * E_{\mathrm{ref}}(t)$. This corresponds to a multiplication in the frequency domain: $\tilde{H}(\omega) \cdot \tilde{E}_{\mathrm{ref}}(\omega)$. As $S_{\mathrm{DFTIR}}(\tau)$ only depends linearly on $E_{\mathrm{sample}}(t)$ in the dispersive setup, the phase information of $\tilde{H}(\omega)$ is obtained after Fourier transform of the interferogram. This is also illustrated by the different nature of the time-domain signals in Fig. 2.6.

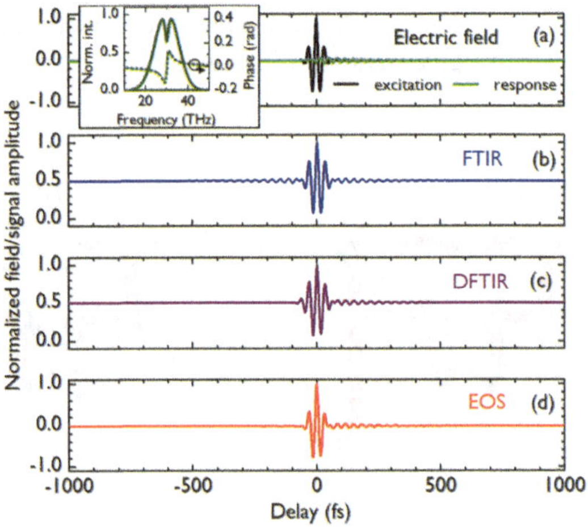

Fig. 2.6 Comparison of signals from FTIR, DFTIR and EOS. **a** MIR electric field (black) and sample response upon excitation with that field (green). The inset shows the corresponding spectrum and spectral phase of the combined field, with a Lorentzian absorption line at 30 THz. **b** Simulated interferogram for an FTIR measurement. **c** Simulated interferogram for a dispersive FTIR measurement. **d** Simulated time-domain EOS trace, for a gate pulse with 5 fs pulse duration and a central wavelength of 1550 nm

Panel (a) shows the time-domain MIR electric field and the response of a sample after excitation with that field. The corresponding complete spectrum and spectral phase are plotted in the inset, showing the Lorentzian absorption line at 30 THz which was used to simulate a sample response. The simulated interferogram for the FTIR measurement in panel (b) is centred around a relative signal strength of 0.5, because of the constant background of the two intensities. Furthermore, the signal is symmetric in time, which directly reflects the loss of phase information. The simulated interferogram of the dispersive FTIR is also offset at 0.5, but the molecular signal only appears in the wake of the 'central burst', reflecting the conserved phase information of the molecular response. For both geometries, the common background can be subtracted using a balanced detection scheme. However, the integrated intensities of the pulses in both paths arrive at the detector at all times, contributing intensity noise. Panel (d) shows a simulated EOS trace, using gate pulses with 5 fs pulse duration at 1550 nm in a 1-μm-thin GaSe detection crystal and no spectral post-filtering. As discussed below (see Fig. 3.7a), the spectral response is flat for those parameters. The simulated trace closely resembles the input MIR field up to a small phase shift. Due to the temporal gating, the intensity noise of the main pulse of the MIR field does not affect the measurement of the molecular signal trailing the few-cycle excitation.

2.4 Nonlinear Crystals and Phasematching

Frequency conversion *via* nonlinear effects relies on suitable materials with fitting symmetries and nonlinear coefficients. Material properties for most nonlinear crystals are summarized in [18, 28]. The choice of the detection crystal material for EOS in the THz spectral region is discussed in [29]. Furthermore, Wilke and Sengupta [30] gives an overview of the most common materials for EOS detection.

Here, several nonlinear crystals which are suitable for MIR detection with EOS, employing the second-order nonlinear processes of difference and sum-frequency generation, are compared. These materials do not have inversion symmetry, in order for the second-order susceptibility to be non-vanishing ($\chi^{(2)} \neq 0$). Key parameters for a comparison of their performance, summarized for all crystals in Table 2.1, are:

1. Nonlinear coefficient ('Nonlin.'): Determines the interaction strength of the fields in the nonlinear process. As mentioned in Sect. 2.3.2 on the Pockels-effect picture, nonlinear crystals for the THz spectral region are typically characterized using the electro-optic coefficient r_{ij} instead of the nonlinear coefficient d_{ij}. The two are related *via* the refractive index of the gate pulse n_{gate} [31]: $d_{\text{ij}} = \frac{1}{4} n_{\text{gate}}^4 r_{\text{ij}}$. In the table below, the nonlinear coefficients for ZnTe ($r_{41} = 3.9$ [30]) and GaP ($r_{41} = 0.97$ [30]) are calculated from their electro-optic coefficients using the refractive indices $n_{\text{ZnTe}} = 2.8$ [32] and $n_{\text{GaP}} = 3.1$ [33] at 1030 nm, respectively.
2. The spectral window of optical transmission ('Transm.').
3. Walk-off: Temporal walk-off between the gate and MIR field, caused by differences in the group refractive index. This factor is a measure for the phase-

Table 2.1 Overview of second-order nonlinear crystals for DFG and EOS

Crystal	Nonlin. (pm/V)	Transm. (μm)	Walk-off (fs)	Bandgap (eV)
ZnTe	60	–	32	2.26 [35]
GaP	22	–	29	2.3 [36]
GaSe	54	0.65–18 [34]	82	2.0–2.09
AGS	16.2 [34]	0.47–13	44	2.7 [37]
ZGP	≈ 70	2–12 [34]	24	2–2.1
HGS	36.8 [34]	0.55–12	43	2.84
CSP	84.5 [38]	0.66–6.5 [34]	97	2.45 [39]
LGS	5.8 [40]	0.32–11.6	7	3.62–4.15
BGSe	9.4–14 [41]	0.47–18 [42]	–	2.73 [43]
LNB	4.6	0.4–5.5	–	3.9–4.3
BBO	2.2	0.2–2.6	–	6.2

mismatch, see the detailed description of EOS in Sect. 2.3 and the calculations in Sect. 3.4. In this table, the temporal walk-off was calculated for the typical MIR and gate pulse wavelengths of 8.6 μm and 1030 nm, respectively, for a 100-μm-thick crystal, using the group refractive indices from [34] for the MIR propagating on the extraordinary crystal axis and the NIR on the ordinary axis.

4. Bandgap for the direct transition: Indicator for the damage threshold by multi-photon absorption, which increases with the bandgap and decreases with the NIR photon energy.

Where not cited differently, the material properties in Table 2.1 are taken from [28].

In general, a nonlinear medium has three refractive indices, one for each of the main crystal axes. The number of independent components depends on the crystal geometry, typically illustrated using the refractive index ellipsoid [17]. In the following, the nonlinear materials listed in Table 2.1 are discussed in more detail, grouped according to the number of independent refractive indices.

Isotropic Crystals

Gallium phosphide (GaP) and zinc telluride (ZnTe) are isotropic zinc blende structure semiconductors with $\bar{4}3$ m point group symmetry. Without an external electric field, all three principle refractive indices are identical. For these materials, the electro-optic effect is described in [9, 21, 44] as field-induced birefringence. Due to the isotropy, rotating the crystals does not change their refractive indices. However, they intrinsically phasematch certain wavelength combinations. GaP, typically cut in [110]-direction, phasematches few-THz radiation and gate pulses with a central wavelength of 1030 nm [36]. This crystal was therefore also chosen for broadband THz generation and electro-optic detection in [24] and subsequently in [45]. High average THz-powers were demonstrated in [46]. A phonon resonance of GaP at 11 THz [47] limits the detection bandwidth at high frequencies. The orientation patterned version of GaP is used to improve the efficiency of difference-frequency generation, e.g. in [48]. ZnTe phasematches few-THz radiation and gate pulses with a central wavelength at 800 nm, e.g., radiation generated by titanium-sapphire oscillators. It is used for THz generation and EOS [49, 50]. The phonon resonance is at 5.31 THz, limiting the detection bandwidth even more than in the case of GaP.

Uniaxial Crystals

Gallium selenide (GaSe) is a negative uniaxial crystal with hexagonal, $\bar{6}2$ m point group symmetry. It has broad phasematching and a high nonlinear coefficient, which is why it is widely used for electro-optic detection of MIR radiation [51]. Silver gallium sulfide (AGS) has similar properties and is employed for EOS detection e.g., in [37]. Zinc germanium phosphide (ZGP) is a new nonlinear material with an outstandingly high nonlinear coefficient of 111 pm/V. However, it suffers from absorption below 2 μm. So far, it was used for DFG with 2.5 μm driving pulses in [52]. It is also promising for electro-optic sampling with long-wavelength gate pulses. One additional new crystal with a high nonlinear coefficient is mercury thiogallate ($HgGa_2S_4$, HGS). It has been used for DFG in [53], its transmission is however limited to 9 μm at long MIR wavelengths. Cadmium silicon phosphide (CSP) is one further negative uniaxial crystal used for MIR generation [38, 39, 54, 55]. It only transmits up to 9 μm and has a drop in transmission to below 60 % at 6.8 μm.

Lithium niobate ($LiNbO_3$, LNB) is used in its periodically poled form (PPLN) to generated MIR radiation up to 5 μm wavelength [56–58]. It is used in [23] for electro-optic detection of low-THz waves and is to be tested for EOS of short-wavelength MIR radiation. Beta barium borate (BBO) is commonly used for second-harmonic generation due to its high bandgap and thus damage threshold. It is used for EOS of near-infrared waveforms [59] and common for second-harmonic generation. The optical transmission is however limited to below 2.5 μm.

Biaxial Crystals

Two bi-axial crystals, commonly used for DFG are promising also for application in EOS: Lithium gallium sulfide ($LiGaS_2$, LGS) is widely used for DFG with 1 μm driving pulses [60]. LGS has the lowest temporal walk-off of all crystals considered here and thus allows to boost the nonlinear conversion by using thick nonlinear media while maintaining a broad spectral coverage. However, its birefringence affects the gate pulse polarization and thus limits the detection noise floor in EOS (see experimental investigation in Sect. 4.1). Furthermore, barium gallium selenide ($BaGa_4Se_7$, BGSe) was recently discovered for DFG [42, 43, 61], with its optical transmission reaching up to 12 μm wavelength.

For the EOS simulations and experiments discussed below, we chose the uni-axial crystal GaSe as the nonlinear crystal, due to its combination of large optical transmission range and nonlinear coefficient. ZGP will be a promising candidate for electro-optic detection with long-wavelength gate pulses, as its transmission only exceeds 50 % at 2 μm wavelength. We excluded the isotropic crystals because their intrinsic phasematching for low-frequency THz radiation is not suitable for the detection of 6-18 μm radiation. The bi-axial crystals are promising because of their remarkably low temporal walk-off, proper handling of gate pulse polarization distortions due to their large birefringence is, however, required. LGS was tested as a detection crystal in the frame of this thesis (Sect. 4.1).

Birefringent Phasematching

Nonlinear optical processes are only efficient if the newly generated radiation adds up coherently upon propagation through the nonlinear material. Assuming no depletion of the incident intensities I_1 and I_2 of the frequency components ω_1 and ω_2, and neglecting dispersion effects, the intensity I_3 of the generated sum-frequency component at frequency $\omega_3 = \omega_2 + \omega_1$ is written as [4]:

$$I_3 = \frac{8 d_{\mathrm{eff}}^2 \omega_3^2 I_1 I_2}{n_1 n_2 n_3 \epsilon_0 c^2} L^2 \mathrm{sinc}\left(\frac{\Delta k L}{2}\right) \tag{2.44}$$

where d_{eff} is the effective nonlinear coefficient [5], n_i are the refractive indices at the interacting frequencies and L the length of the nonlinear medium. The phase-mismatch Δk is defined as

$$\Delta k = k_3 - (k_2 + k_1) = \frac{n_3 \omega_3 - (n_2 \omega_2 + n_1 \omega_1)}{c} \tag{2.45}$$

Due to dispersion, $\Delta k = 0$ is usually not fulfilled, limiting the scaling of the generated sum-frequency power with the square of the interaction length. One method to achieve phasematching is to use birefringent materials. For example, uni-axial materials which have one optical axis have two different main refractive indices, n_o and n_{extra}. Polarization components parallel to the optical axis propagate as the extra-ordinary beam with refractive index n_{extra}, those with perpendicular polarization as the ordinary one with refractive index n_o. The extraordinary refractive index depends on the angle θ between the propagation axis of the light field and the optical axis of the crystal. The tuning angles of a uniaxial nonlinear material are illustrated in Fig. 2.3a. For GaSe, which cannot be cut along a certain θ-angle, the crystal itself has to be rotated. The angle φ between the plane spanned by the propagation direction and the optical axis of the crystal to its x-axis defines the strength of the individual second-order processes that can occur in the nonlinear material, see the rotation matrix U in the previous section. The θ-dependent extra-ordinary refractive index is calculated according to

$$\frac{1}{n_{\mathrm{e}}(\theta)^2} = \frac{\sin^2(\theta)}{n_{\mathrm{extra}}^2} + \frac{\cos^2(\theta)}{n_{\mathrm{o}}^2} \tag{2.46}$$

Figure 2.3b shows the refractive indices for GaSe, with the extraordinary refractive index for several internal (after refraction) values θ_{int}. To achieve phasematching for sum- and difference-frequency generation in EOS, we chose the configuration where the gate pulses propagate on the ordinary crystal axis while the MIR and SFG/DFG components propagate on the extra-ordinary one. Comparing the refractive indices at the near-infrared wavelengths between 1 and 2 μm, which are the crucial components in the equation for the phase-mismatch because of their higher frequencies, an increase of the slopes of the refractive indices is observed for shorter wavelengths. This dispersion behaviour indicates better phasematching for longer gate pulse wavelengths. Furthermore, with increasing θ_{int}, wavelengths with a larger difference have the same n_o and n_e. Therefore, steeper angles correspond to a better phasematching and thus detection efficiency for higher frequency/shorter wavelength MIR radiation.

2.5 Noise in Balanced Detection

As sketched in Fig. 2.2a, the electro-optic signal is measured as the difference current of two identical photodiodes. This balanced detection setup allows for the efficient suppression of common intensity noise in the two channels [62]. For perfect subtraction of the two signals, the noise floor is either limited by the detector noise, or, for sufficiently high optical powers, by the shot noise. The latter has its origin in the quantum nature of light, meaning that for a given optical power P, only an integer number of photons at wavelength λ exists. The photon number N follows a Poissonian distribution, with width $\sqrt{N}$ [63]. The single-sided power spectral density of the shot noise level is radio-frequency independent and calculated as [64]

$$s_q = \sqrt{\frac{2 \times hc}{P\lambda}} = 1.9 \times 10^{-8}\text{Hz}^{-1/2} \times \sqrt{\frac{1\,\text{mW}}{P}} \times \sqrt{\frac{1064\,\text{nm}}{\lambda}} \tag{2.47}$$

thus decreasing with the optical power and wavelength of the radiation. When using two balanced diodes, only half the power $P/2$ impinges on each of those. In addition, due to the uncorrelated noise of the two diodes, the power spectral density of the balanced noise floor is a factor of $\sqrt{2}$ higher than for a single diode measuring $P/2$.

The gate-pulse shot-noise-balanced noise floor in electro-optic detection is thus proportional to the square root of the number of local-oscillator photons N_{LO}:

$$\sigma_{SN} \propto \sqrt{N_{LO}} \tag{2.48}$$

while the signal strength is proportional to the product of the gate pulse and sum-frequency fields (see Eq. 2.36) and thus the square-roots of the photon numbers:

$$S \propto \sqrt{N_{LO}}\sqrt{N_{SFG}} \tag{2.49}$$

resulting in a scaling of the signal to noise ratio (SNR) or what will be called *dynamic range* in the following, with $\sqrt{N_{SFG}}$ and SNR$=1$ for a single sum-frequency photon.

2.6 Techniques to Improve the EOS Dynamic Range and Sensitivity

Spectral Filtering

As can be seen from the sketch in Fig. 2.2b, for MIR frequencies larger than the spectral width of the gate pulse spectrum, the spectral shift of the generated sum-frequency spectrum is significant. Therefore, the two spectra do not overlap completely, limiting the spectral region of polarization rotation as discussed above. As a consequence, a significant number of gate pulse photons only contributes to noise on the detector, but not to the signal. In [10], improving the detection dynamic range by using a spectral filter to select only the spectral region of interest was demonstrated. For a SPF with cut-on frequency ν_{SPF}, the EOS signal is proportional to:

$$S_{\mathrm{SPF}} \propto \int_{\nu_{\mathrm{SPF}}}^{\infty} A_{\mathrm{LO}} \cdot A_{\mathrm{SFG}} \cdot \nu d\nu \tag{2.50}$$

Assuming the local oscillator to be shot-noise limited, the noise floor (for no MIR radiation) is proportional to:

$$SN_{\mathrm{SPF}} \propto \sqrt{I_{\mathrm{LO}}} \tag{2.51}$$

with the spectrally filtered intensities I_{LO} of the local oscillator. The resulting signal to noise ratio (or what is called dynamic range throughout this thesis) has a maximum for the optimum ν_{SPF}, that depends on the power distribution of the gate and sum-frequency spectra. This is illustrated for the experimental spectra measured in Sects. 4.1 and 4.8 in Fig. 2.7.

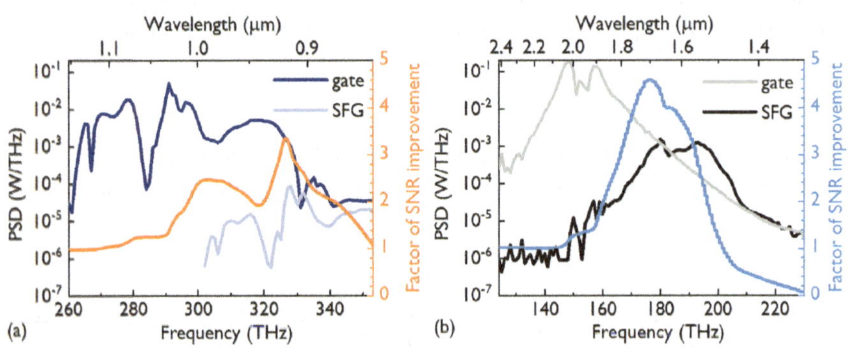

Fig. 2.7 Calculated shot-noise limited SNR improvement as a function of the SPF wavelength for two experimental configurations. **a** Gate pulse and sum-frequency spectrum at the delay position of maximum signal measured using 1030-nm gate pulses. **b** Gate pulse and sum-frequency spectrum at the delay position of maximum signal measured using 1960-nm gate pulses

Panel (a) shows the measured gate pulse and sum-frequency spectrum for the 1030-nm gate-pulse EOS discussed in detail in Sect. 4.1. The calculated improvement of the dynamic range/signal-to-noise ratio as a function of the filter wavelength is shown on the right axis. As the sum-frequency power is below the local oscillator for almost all wavelengths, the optimum cut-off wavelength is at the edge of the spectrum. For the higher sum-frequency powers in the 1960-nm gate-pulse EOS shown in panel (b) and discussed in detail in Sect. 4.8, the optimum cut-off is significantly closer to the central wavelength. The calculated optima will be compared to the filters used in the experiment below.

Selective Attenuation

For a measurement of the electro-optic signal as the ratio of the difference current to the total current from both diodes, attenuation of the local oscillator increases the signal strength as it effectively enhances the measured polarization change. In our case, where we measure the non-normalized difference current, attenuation of the local oscillator can be required to avoid saturation of the balanced diodes, while having no effect on the measured signal to noise ratio as the shot-noise limited noise floor is decreased by the same amount. However, this selective attenuation has to be achieved without affecting the sum-frequency beam. In [65], polarization-selective attenuation using Brewster plates is suggested. Furthermore, using crossed polarizers to reduce the local oscillator power or what is called the 'optical bias' was employed to detect weak THz signals in [66–68] and is discussed in [69].

However, for the comparatively weak local oscillator, potential nonlinearities of the signal with the THz field strength have to be taken into account, as discussed in [26, 70]. For an arbitrary optical bias Γ_0 and no contributions from scattered light, they write the measured signal strength as:

$$I \propto I_0 \cdot \sin^2(\Gamma_0 + \Gamma_{\mathrm{THz}}) \tag{2.52}$$

When using a quarter-wave plate in the standard-geometry, $\Gamma_0 = \frac{\pi}{4}$ and the normalized difference-current simplifies to Eq. (2.41) for the Pockels effect. Assuming however a small optical bias $\Gamma_0 \ll 1$ that is on the same order as the polarization rotation caused by the THz signal, $\Gamma_{\mathrm{THz}} \ll 1$, the signal is approximated as

$$I \approx I_0 \cdot (\Gamma_0 + \Gamma_{\mathrm{THz}})^2 \propto \Gamma_{\mathrm{THz}}^2 \tag{2.53}$$

Considering the THz-induced phase shift in Eq. (2.40), the signal therefore depends on the square of the THz-field. In [26, 70], ways to retrieve the THz-field from the nonlinear measurements are presented.

While we typically do not operate the EOS in the low-bias limit, the generated sum-frequency power is similar to the strength of the local oscillator or even higher. Because we measure the non-normalized signal, the stronger sum-frequency field itself does not result in any nonlinearities, as will be confirmed experimentally in Sect. 4.9.

References

1. Udem T, Holzwarth R, Hänsch TW (2002) Optical frequency metrology. Nature 416:233–237. ISSN: 1476-4687
2. Ye J, Cundiff ST (2005) Femtosecond optical frequency comb: principle, operation, and applications. Springer Science + Business Media Inc, Boston. ISBN: 0387237917
3. Jones RC (1941) A new calculus for the treatment of optical systems description and discussion of the calculus. J Opt Soc Am 31:488. ISSN: 0030-3941
4. Boyd RW (2008) Nonlinear optics, 3rd edn. Academic Press, Burlington
5. Midwinter JE, Warner J (1965) The effects of phase matching method and of uniaxial crystal symmetry on the polar distribution of second-order non-linear optical polarization. Br J Appl Phys 16:1135–1142. ISSN: 0508-3443
6. Brabec T, Krausz F (2000) Intense few-cycle laser fields: frontiers of nonlinear optics. Rev Modern Phys 72:545–591. ISSN: 0034-6861
7. Brabec T, Krausz F (1997) Nonlinear optical pulse propagation in the single-cycle regime. Phys Rev Lett 78:3282. ISSN: 0031-9007
8. Wu Q, Zhang X-C (1995) Free-space electro-optic sampling of terahertz beams. Appl Phys Lett 67:3523–3525. ISSN: 0003-6951
9. Namba S (1961) Electro-optical effect of zincblende. J Opt Soc Am 51:76. ISSN: 0030-3941
10. Porer M, Ménard J-M, Huber R (2014) Shot noise reduced terahertz detection via spectrally postfiltered electro-optic sampling. Opt Lett 39:2435–2438. ISSN: 1539-4794
11. Gallot G, Grischkowsky D (1999) Electro-optic detection of terahertz radiation. J Opt Soc Am B 16:1204. ISSN: 0740-3224
12. Riek C, Seletskiy DV, Leitenstorfer A (2017) Femtosecond measurements of electric fields: from classical amplitudes to quantum fluctuations. Eur J Phys 38:024003. ISSN: 0143-0807
13. Sulzer P et al (2020) Determination of the electric field and its Hilbert transform in femtosecond electro-optic sampling. Phys Rev A 101:033821. ISSN: 1050-2947
14. Leitenstorfer A, Hunsche S, Shah J, Nuss MC, Knox WH (1999) Detectors and sources for ultrabroadband electro-optic sampling: experiment and theory. Appl Phys Lett 74:1516–1518. ISSN: 0003-6951
15. Conforti M, Baronio F, de Angelis C (2011) Modeling of ultrabroadband and singlecycle phenomena in anisotropic quadratic crystals. J Opt Soc Am B 28:1231. ISSN: 0740-3224
16. Roberts DA (1992) Simplified characterization of uniaxial and biaxial nonlinear optical crystals: a plea for standardization of nomenclature and conventions. IEEE J Quant Electron 28:2057–2074. ISSN: 0018-9197
17. Sutherland RL (2003) Handbook of nonlinear optics, 2nd edn. revised and expanded. Marcel Dekker, New York. ISBN: 0-8247-4243-5
18. Dmitriev VG, Gurzadyann GG, Nikogosyan DN (2011) Handbook of nonlinear optical crystals, new. Springer, Berlin. ISBN: 978-3-642-08472-0
19. Tomasino A et al (2013) Wideband THz time domain spectroscopy based on optical rectification and electro-optic sampling. Sci Rep 3:3116
20. Chen Q, Tani M, Jiang Z, Zhang X-C (2001) Electro-optic transceivers for terahertzwave applications. J Opt Soc Am B 18:823. ISSN: 0740-3224
21. Planken PCM, Nienhuys H-K, Bakker HJ, Wenckebach T (2001) Measurement and calculation of the orientation dependence of terahertz pulse detection in ZnTe. J Opt Soc Am B 18:313. ISSN: 0740-3224
22. Oguchi K, Yasumatsu N, Watanabe S (2014) Polarization detection of terahertz radiation via the electro-optic effect using zinc blende crystal symmetry. J Opt Soc Am B31:3170. ISSN: 0740-3224
23. Oguchi K, Iwasaki H, Okano M, Watanabe S (2016) Polarization-sensitive electrooptic detection of terahertz wave using three different types of crystal symmetry: toward broadband polarization spectroscopy. Appl Phys Lett 108:011105. ISSN: 0003-6951
24. Xu J et al (2018) Three-octave terahertz pulses from optical rectification of 20 fs, 1 μm, 78 MHz pulses in GaP. J Phys B: Atom Molecul Opt Phys 51:154002. ISSN: 0953-4075

25. Yariv A (1989) Quantum electronics, 3rd edn. Wiley, New York. ISBN: 0471609978
26. Brunner FDJ et al (2014) Distortion-free enhancement of terahertz signals measured by electro-optic sampling. I. Theory. J Opt Soc Am B 31:904. ISSN: 0740-3224
27. Parker TJ (1990) Dispersive fourier transform spectroscopy. Contemp Phys 31:335–353. ISSN: 0010-7514
28. Nikogosyan DN (2005) Nonlinear optical crystals: a complete survey. Springer-Science, New York. ISBN: 0-387-22022-4
29. Pradarutti B et al (2008) Highly efficient terahertz electro-optic sampling by material optimization at 1060 nm. Opt Commun 281:5031–5035. ISSN: 00304018
30. Wilke I, Sengupta S (2008). In: Dexheimer SL (ed) Terahertz spectroscopy–principles and applications. CRC, Boca Raton, pp 41–72
31. Boyd GD, Pollack MA (1973) Microwave nonlinearities in anisotropic dielectrics and their relation to optical and electro-optical nonlinearities. Phys Rev B 7:5345. ISSN: 0163-1829
32. Li HH (1984) Refractive index of ZnS, ZnSe, and ZnTe and its wavelength and temperature derivatives. J Phys Chem Ref Data 13:103–150. ISSN: 0047-2689
33. Bond WL (1965) Measurement of the refractive indices of several crystals. J Appl Phys 36:1674–1677. ISSN: 0021-8979
34. SNLO nonlinear optics code available from A. V. Smith, AS-Photonics, Albuquerque, NM
35. Rumble JR, Bruno TJ (eds) (2020) CRC handbook of chemistry and physics: a ready-reference book of chemical and physical data 2020–2021, 101st edn. CRC Press, Boca Raton. ISBN: 0367417243
36. Wu Q, Zhang X-C (1997) 7 terahertz broadband GaP electro-optic sensor. Appl Phys Lett 70:1784–1786. ISSN: 0003-6951
37. Riek C et al (2015) Direct sampling of electric-field vacuum fluctuations. Science 350:420–423. ISSN: 1095-9203
38. Chaitanya Kumar S et al (2015) High-power femtosecond mid-infrared optical parametric oscillator at 7 μm based on CdSiP2. Opt Lett 40:1398. ISSN: 1539-4794
39. Zawilski KT et al (2010) Growth and characterization of large CdSiP2 single crystals. J Cryst Growth 312:1127–1132. ISSN: 0022-0248
40. Petrov V et al (2004) Second harmonic generation and optical parametric amplification in the mid-IR with orthorhombic biaxial crystals LiGaS2 and LiGaSe2. Appl Phys B 78:543–546. ISSN: 0946-2171
41. Boursier E et al (2015) Angle noncritical phase-matched second-harmonic generation in the monoclinic crystal BaGa4Se7. Opt Lett 40:4591–4594. ISSN: 1539-4794
42. Boursier E et al (2016) Phase-matching directions and refined Sellmeier equations of the monoclinic acentric crystal BaGa4Se7. Opt Lett 41:2731–2734. ISSN: 1539-4794
43. Yelisseyev AP, Lobanov SI, Krinitsin PG, Isaenko LI (2020) The optical properties of the nonlinear crystal BaGa4Se7. Opt Mater 99:109564. ISSN: 0925-3467
44. Moskalenko AS, Riek C, Seletskiy DV, Burkard G, Leitenstorfer A (2015) Paraxial theory of direct electro-optic sampling of the quantum vacuum. Phys Rev Lett 115. ISSN: 0031-9007
45. Barbiero G et al (2020) Broadband terahertz solid-state emitter driven by Yb:YAG thindisk oscillator. J Phys B: Atom Molecul Opt Phys 53:125601. ISSN: 0953-4075
46. Meyer F et al (2019) Milliwatt-class broadband THz source driven by a 112 W, sub-100 fs thin-disk laser. Opt Exp 27:30340–30349
47. Wu Q, Zhang X-C (1996) Design and characterization of traveling-wave electrooptic terahertz sensors. IEEE J Select Top Quant Electron 2:693–700. ISSN: 1558-4542
48. Lee KF, Hensley CJ, Schunemann PG, Fermann ME (2017) Midinfrared frequency comb by difference frequency of erbium and thulium fiber lasers in orientationpatterned gallium phosphide. Opt Exp 25:17411–17416. ISSN: 1094-4087 (2017)
49. Wu Q, Litz M, Zhang X-C (1996) Broadband detection capability of ZnTe electro-optic field detectors. Appl Phys Lett 68:2924–2926. ISSN: 0003-6951
50. Nahata A, Weling AS, Heinz TF (1996) A wideband coherent terahertz spectroscopy system using optical rectification and electro-optic sampling. Appl Phys Lett 69:2321–2323. ISSN: 0003-6951

51. Liu K, Xu J, Zhang X-C (2004) GaSe crystals for broadband terahertz wave detection. Appl Phys Lett 85:863–865. ISSN: 0003-6951
52. Vasilyev S et al (2019) Super-octave longwave mid-infrared coherent transients produced by optical rectification of few-cycle 25-μm pulses. Optica 6:111. ISSN: 2334-2536
53. Beutler M, Rimke I, Büttner E, Badikov V, Petrov V (2016) Femtosecond mid-IR difference-frequency generation in HgGa2S4 from an 80 MHz optical parametric oscillator pumped at 800 nm. J Opt Soc Am B 33:D7. ISSN: 0740-3224
54. Kemlin V et al (2011) Nonlinear, dispersive, and phase-matching properties of the new chalcopyrite CdSiP2. Opt Mater Exp 1:1292. ISSN: 2159-3930
55. Sánchez D et al (2014) Broadband mid-IR frequency comb with CdSiP2 and AgGaS2 from an Er,Tm:Ho fiber laser. Opt Lett 39:6883. ISSN: 1539-4794
56. Seidel M et al (2018) Multi-watt, multi-octave, mid-infrared femtosecond source. Sci Adv 4:eaaq1526
57. Lind AJ et al (2020) Mid-infrared frequency comb generation and spectroscopy with few-cycle pulses and chi(2) nonlinear optics. Phys Rev Lett 124:133904. ISSN: 0031-9007
58. Catanese A et al (2020) Mid-infrared frequency comb with 6.7 W average power based on difference frequency generation. Opt Lett 45:1248-1251. ISSN: 1539-4794
59. Keiber S et al (2016) Electro-optic sampling of near-infrared waveforms. Nat Photon 10:159–162. ISSN: 1749-4885
60. Pupeza I et al (2015) High-power sub-two-cycle mid-infrared pulses at 100 MHz repetition rate. Nat Photon 9:721–724. ISSN: 1749-4885
61. Zhang J et al (2020) Broadband, few-cycle mid-infrared continuum based on the intrapulse difference frequency generation with BGSe crystals. Opt Exp 28:37903–37909
62. Hobbs PCD (1997) Ultrasensitive laser measurements without tears. Appl Opt 36:903. ISSN: 0003-6935
63. Kingston RH (1979) Detection of optical and infrared radiation. Springer, Berlin. ISBN: 9780387086170
64. Kwee P, Willke B, Danzmann K (2011) New concepts and results in laser power stabilization. Appl Phys B 102:515–522. ISSN: 0946-2171
65. Ahmed S, Savolainen J, Hamm P (2014) Detectivity enhancement in THz electrooptical sampling. Rev Sci Instrum 85:013114. ISSN: 1089-7623
66. Jiang Z, Sun FG, Chen Q, Zhang X-C (1999) Electro-optic sampling near zero optical transmission point. Appl Phys Lett 74:1191–1193. ISSN: 0003-6951
67. Chen Q, Zhang X-C (1999) Polarization modulation in optoelectronic generation and detection of terahertz beams. Appl Phys Lett 74:3435–3437. ISSN: 0003-6951
68. Degert J, Cornet M, Abraham E, Freysz E (2016) Simple and distortion-free optical ampling of terahertz pulses via heterodyne detection schemes. JOSA B 33:2045-2050. ISSN: 1520-8540
69. Krauspe P, Banerji N, Réhault J (2020) Effective detection of weak terahertz pulses in electro-optic sampling at kilohertz repetition rate. JOSA B 37:127–132. ISSN: 1520-8540
70. Johnson JA et al (2014) Distortion-free enhancement of terahertz signals measured by electro-optic sampling. II. experiment. J Opt Soc Am B 31:1035. ISSN: 0740-3224

Chapter 3
Simulation Results for EOS

For an in-depth understanding of EOS-based detection of MIR waves, a one-dimensional numerical model was developed, based on the theoretical description in Sect. 2.3. It considers sum- and difference-frequency generation in GaSe and the subsequent heterodyne detection with an ellipsometry setup using a quarter-wave plate and Wollaston prism. The nonlinear propagation according to the first-order propagation equation is solved in a split-step method: the dispersion of the interacting fields is modelled in the frequency domain, while the nonlinear polarization results from the time-domain multiplication of the fields. As the full second-order polarization is considered, phasematching of sum- and difference-frequency generation as well as cascaded mixing effects are taken into account.

Using this model, several questions were investigated and are discussed in the following: First, the effect of the spectral phase of the gate pulse and of propagation through thick nonlinear crystals is illustrated by simulating the wavelength-dependent electro-optic signal. In the second section, differences in the detection efficiency between phasematching SFG and DFG in EOS are investigated. For a reconstruction of the electric field from the measured trace, the frequency-domain instrument response function and its time-domain correspondent, the effective temporal gate window are introduced in the third section. In the fourth section, the spectral instrument response is compared for gate pulses with varying central wavelength. This will also be explored experimentally in the next chapter. Finally, the dependence of the instrument response function (IRF) on the spectral phase of the gate pulse is simulated for 11-fs gate pulses at 1550 nm central wavelength.

© The Author(s), under exclusive license to Springer Nature Switzerland AG 2022

C. Hofer, *Detection Efficiency and Bandwidth Optimized Electro-Optic Sampling of Mid-Infrared Waves*, Springer Theses,
https://doi.org/10.1007/978-3-031-15328-0_3

3.1 Wavelength-Dependent EOS Signal for Thick Detection Crystals

As described in detail in [1], by investigating the wavelength-dependent EOS signal, the ellipsometry setup measures the relative phase between the local oscillator and the generated sum- and difference-frequency components, which includes the phase of the measured field. A straightforward expression for this relative phase was given for thin detection crystals and a compressed gate pulse. Using a quarter-wave plate in the ellipsometry setup, the relative phase is constant for all NIR wavelengths and the signals from the different wavelengths add up constructively.

In this work, >50-μm-thick detection crystals are used to maximize the MIR-to-NIR conversion efficiency. However, their dispersion and polarization dependent propagation effects cause the relative phase between local oscillator and sum-frequency radiation to be wavelength dependent, potentially reducing the overall EOS signal strength. Therefore, the effect of the spectral phase of the gate pulse and of the crystal dispersion on the wavelength dependent EOS signal was analyzed with simulations.

The input fields were gate pulses with a central wavelength of 1550 nm and a full-width-at-half-maximum (FWHM) pulse duration of 11 fs (see Fig. 4.19 for the measured pulse shape) and a Fourier-transform limited, super-Gaussian MIR spectrum with a central wavelength of 10 μm and a spectral FWHM of 6 THz. First, different gate pulse phases are compared in Fig. 3.1 for a 1-μm-thick GaSe crystal, where dispersion is negligible.

The top-row depicts the delay and NIR-frequency-dependent spectral intensity of the generated sum- and difference-frequency components. In the middle row, the input field and the EOS signal are compared and the bottom row shows the wavelength-dependent EOS signal. From left to right (panel (a) to (c)), the spectral phase of the gate pulse was varied: To the left, the experimentally available gate pulse was taken as measured using FROG [2], which has a strongly varying spectral phase below 1200 nm (250 THz). In panel (b), the spectral phase of the gate pulse was set to zero, thus resulting in a Fourier transform limited pulse, which closely resembles the ideal case described in [1]. To the right, the spectral phase of 3-mm-thick fused silica was added to the Fourier-transform limited pulse. For all gate pulses, the spectral intensity of the sum- and difference-frequency components shows a periodic modulation with twice the frequency of the detected field, which allows for a determination of the square of the field amplitude [3]. This is caused by interference of the two contributions, that are generated with equal strength in the thin crystal, with their relative phase varying with the gate pulse delay. No fringes are visible in the wings of the spectra, where only one of the processes occurs. For the very thin detection crystal, the spectral phase of the gate pulse is directly reflected in the delay-dependent shape of the generated sum- and difference-frequency components [4]. The wavelength-dependent EOS signal shows a similar behaviour to the delay-dependent sum- and difference-frequency spectra. When integrating over all wavelengths to achieve the simulated EOS traces, contributions with opposite sign cancel out and reduce the

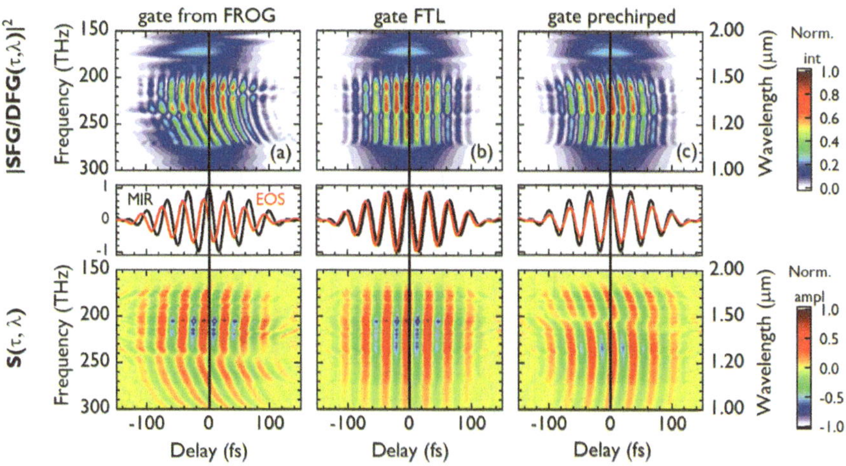

Fig. 3.1 Influence of the spectral phase of the gate pulse on the EOS signal for a 1-µm-thick detection crystal. Top row: Sum- and difference-frequency intensity as a function of the gate pulse delay and frequency. Middle: comparison of the input field and the EOS trace. Bottom: Frequency and delay dependent EOS signal. **a** Gate pulse from FROG measurement. **b** Fourier transform limited gate pulse, $\varphi(\omega) = 0$. **c** The phase of 3-mm-thick fused silica added to a Fourier transform limited gate pulse

signal strength. In addition, the reduced peak power and a longer gate-time window have detrimental effects on the signal strength. This reduction is visible in the comparison of the relative strengths of the EOS traces in the middle row, which is the lowest for the gate pulse taken from the FROG measurement and the strongest for the Fourier-transform limited gate pulse. In this case, the simulated trace resembles the input field up to a small time-shift. Benefits from pre-chirping with fused silica for thicker detection crystals will be explained below.

As discussed in Sect. 2.4, the intensity of the generated sum- and difference-frequency components scales with the square of the length of the detection crystal. However, dispersion and temporal walk-off of the interacting fields limit this behaviour. Here, the wavelength-dependent EOS signal was calculated for a 30-µm-thick and a 100-µm-thick detection crystal with the same MIR waveform and gate pulse parameters as for the previous figure, to illustrate the effect of the increasing crystal thickness on the relative phase between the generated sum-frequency components and the gate pulse. With otherwise identical input parameters as in Fig. 3.1, this is illustrated in Fig. 3.2.

Again, the top row shows the delay-dependent intensity of the generated sum- and difference-frequency components, the middle row the input fields and EOS traces and the wavelength-dependent signal is illustrated in the bottom row. In panels (a) and (b), a 30-µm and a 100-µm-thick detection crystal were used, respectively. Panel (c) shows the results for a 100-µm-thick crystal, where the spectral phase of 4-mm-thick fused silica was added to the gate pulse. The GDD of fused silica of this thickness

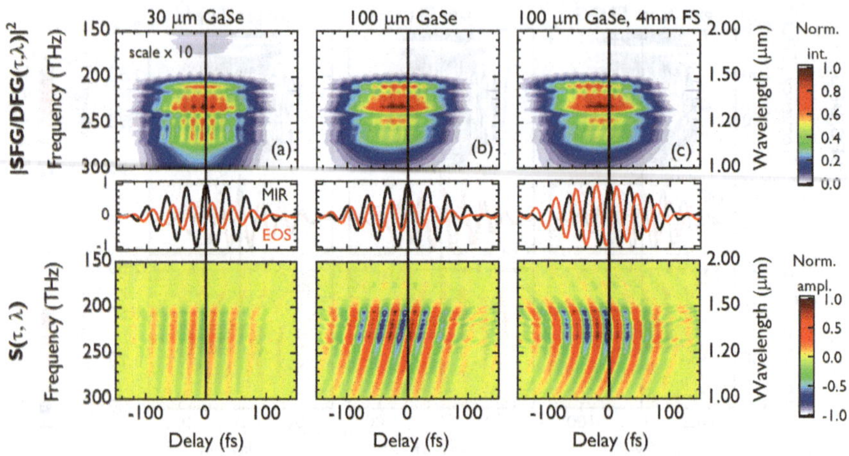

Fig. 3.2 Frequency-dependent EOS signal for thick detection crystals. Top row: Sum- and difference-frequency intensity as a function of the gate pulse delay and frequency. Middle: comparison of the input field and the EOS trace. Bottom: Frequency- and delay-dependent EOS signal. **a** 30-μm-thick GaSe and Fourier-transform-limited gate pulse. **b** 100-μm-thick GaSe and Fourier-transform-limited gate pulse. **c** 100-μm-thick GaSe and gate pulse pre-chirped with 4-mm-thick fused silica

matches that of 50-μm of GaSe at $\approx$ 1450 nm but with opposite sign, resulting in temporal compression of the gate pulse at this wavelength in the center of the crystal.

For the increased crystal thicknesses, the generated sum- and difference-frequency spectra show less modulation with delay. This is caused by a preference of SFG with phasematching in the thicker crystals that reduces the DFG contributions and thus the modulation depth. An increase of the slope of the fringes of the wavelength dependent signal is noted when increasing the crystal thickness (compare (a) and (b)). Prechirping the gate pulse with fused silica leads to regions with vertical fringes, where the spectral phase of fused silica flattens the phase relation between the generated sum-frequency components and the local oscillator. The limited spectral region of a constant relative phase decreases the signal strength when integrating over the full wavelength range. Pre-chirping the gate pulse thus has two effects when using thick detection crystals: First, instead of temporal elongation of the gate pulse through the whole crystal when entering perfectly compressed, the point of optimum compression can be shifted to the center of the crystal. On average, this increases the achievable peak powers in the crystal and thus the efficiency of the sum-frequency generation. Secondly, compensating for the spectral phase of the detection crystal also causes the phase difference between the gate pulse and local oscillator to be constant over broader spectral regions and thus increases the spectrally integrated signal strength.

3.2 MIR Depletion and Amplification for Phasematching SFG or DFG

In thin nonlinear crystals with broadband phasematching, SFG and DFG are equally strong, as illustrated in Fig. 3.1 by the interference of the two components. When employing thicker detection crystals, e.g., in order to maximize the detection efficiency, one of the components is preferred, depending on the phasematching configuration. In the experiments shown below, this is typically the sum-frequency component. However, SFG and DFG are fundamentally different in photon rates. The level schemes of both processes are illustrated in Fig. 3.3 [5].

In SFG (left), two input photons at frequencies ω_1 and ω_2 are converted into one output photon at the sum of the two frequencies, $\omega_3 = \omega_1 + \omega_2$. On the other hand, in the DFG process (right), the mixing of the input frequencies generates an output photon at the difference of the two frequencies, $\omega_3 = \omega_1 - \omega_2$. For energy conservation, a second photon at ω_2 is generated in this process. Therefore, the investigated field is amplified during this process.

For optimizing the electro-optic detection efficiency, one step is the maximization of the conversion of MIR photons to the NIR. When phasematching SFG, the MIR spectrum can be depleted significantly, as shown experimentally in Sect. 4.8, potentially causing a saturation of the EOS signal strength for thick crystals. Because of the fundamentally different nature of the DFG process (see Fig. 3.3), no depletion, but instead an amplification of the investigated field occurs when phasematching this contribution. Therefore, potential detection efficiency advantages of phasematching DFG instead of SFG were investigated theoretically.

To this end, detection of a super-Gaussian MIR spectrum of 4.5 μm width, centred at 10 μm and with 1 mW average power was simulated for two different phasematching configurations, both with perpendicular polarizations for the gate and MIR pulses,

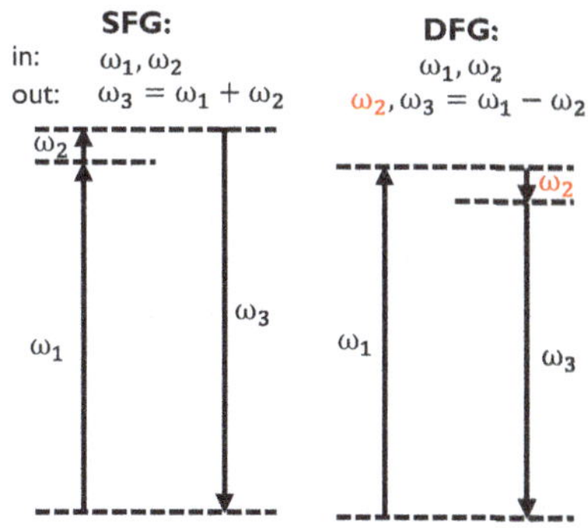

Fig. 3.3 Level schemes of the second-order nonlinear effects of SFG (left) and DFG (right). In SFG two input photons at frequencies ω_1 and ω_2 mix to generate one photon at frequency $\omega_3 = \omega_1 + \omega_2$. In DFG, two input photons at frequencies ω_1 and ω_2 mix to generate a photon at frequency $\omega_3 = \omega_1 - \omega_2$. For energy conservation, an additional photon at frequency ω_2 is generated

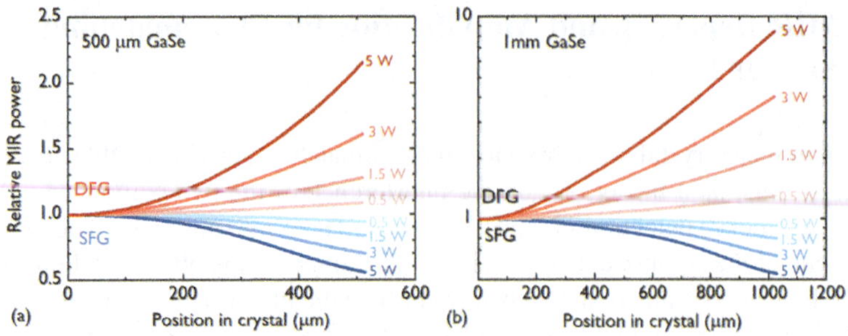

Fig. 3.4 MIR power evolution through the EOS crystal at the gate pulse delay of maximum signal for different gate pulse powers and crystal thicknesses. **a** 500-μm-thick EOS crystal. **b** 1-mm-thick EOS crystal

as this is the case in our experiments. For SFG, the phasematching angles were set to $\theta = 12.9°$ and $\varphi = 0°$, with the gate pulse propagating on the ordinary and the MIR and sum-frequency on the extra-ordinary polarization axis. For DFG, $\theta = 11.1°$ and $\varphi = 30°$, with opposite polarization axes as for SFG. In both cases, the gate pulse power was varied from 0.5 to 5 W, for a repetition rate of 50 MHz and a beam radius of 35 μm, thus increasing the nonlinear conversion efficiency. Their central wavelength was 1960 nm and the FWHM pulse duration was 13 fs, with a temporal shape similar to that of the FROG measurements shown in Fig. 4.29. Figure 3.4 illustrates the change of the MIR average power through the detection crystal at the gate pulse delay position of maximum EOS signal for the two phasematching configurations.

In panel (a), the relative MIR power evolution is shown on a linear scale for a 500-μm-thick detection crystal. When phasematching SFG, the final power value at the end of the detection crystal monotonically decreases with increasing gate pulse power, reaching as low as 50 % for 5 W. For phasematching DFG, the average infrared output power increases with the gate pulse power and is doubled for the highest one. For the 1-mm-thick crystal, the power evolutions is shown on a logarithmic scale in panel (b). Similar effects to the thinner crystal are observed. However, for SFG and 5 W gate pulse power, the MIR power reaches a minimum before the end of the crystal, while the amplification does not saturate in the case of DFG. For these power levels, advantages from phasematching DFG could become significant. This is further investigated, by plotting the sum- and difference-frequency spectra and the spectrally resolved MIR depletion at the peak of the EOS signal, see Fig. 3.5.

Panels (a) and (c) show the generated sum- and difference-frequency spectra for a 500-μm and 1-mm-thick crystal, respectively. The black dashed line indicates the gate pulse/local oscillator spectrum for comparison. In panels (b) and (d), the corresponding MIR spectra at the end of the detection crystal are depicted. With increasing gate pulse power, the level of the sum-frequency spectra (blue) starts to saturate, while a monotonic increase is observed for the generated DFG spectra.

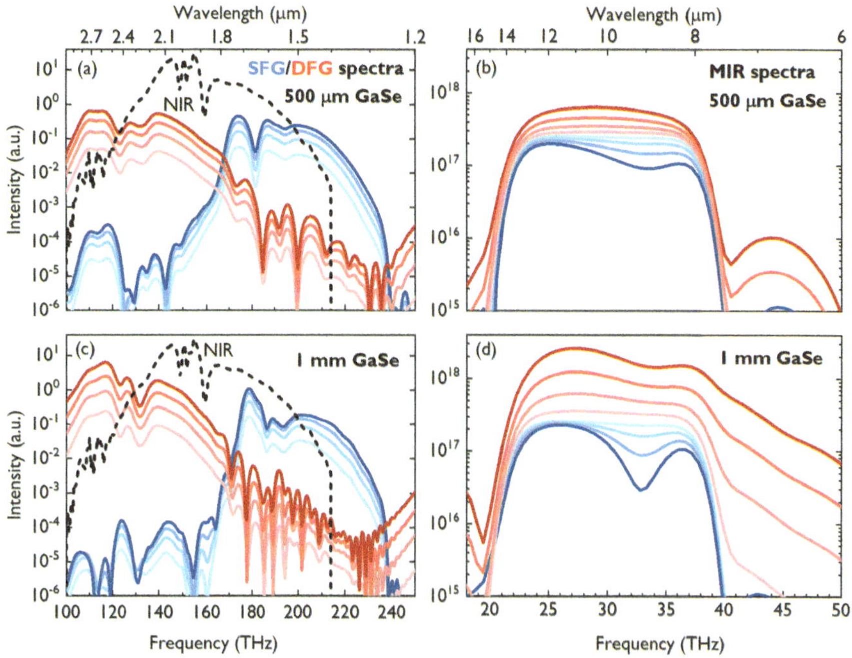

Fig. 3.5 **a**, **c** Simulated sum- (blue) and difference-frequency (red) spectra at maximum EOS signal for 500-μm-thick and 1-mm-thick GaSe, respectively; Gate pulse spectrum (dashed line) shown for comparison. **b**, **d** Corresponding MIR spectra at the end of the EOS crystal for the delay value of maximum signal strength

Furthermore, in the case of the 1-mm-thick crystal, the generated spectra are more narrowband than for the 500-μm-thick one. This effect is explained by the reduced phasematching bandwidth with increasing crystal thickness. Similarly, the spectral depletion increases for SFG, both with the gate pulse power and with the crystal thickness. For the 1-mm-thick crystal, this spectral dip is also more narrowband. In the case of DFG, the MIR amplification is spectrally flat for the 500-μm-thick crystal, while it increases at longer wavelengths for the thicker crystal. Furthermore, the generation of short-wavelength infrared radiation increases with the crystal thickness, most likely due to cascaded effects with the newly generated NIR photons at longer wavelengths or second harmonic generation of the generated MIR.

For the gate pulse power of 1.5 W which is at the onset of saturation for SFG and close to the experimental value (1.9 W), the simulated EOS spectra are compared in Fig. 3.6.

For phasematching SFG, the EOS spectrum becomes more narrowband but also locally stronger for the thicker crystal. In the case of DFG, this effect is not observable, because the MIR amplification between 11 and 14 μm dominates the signal. As no spatial effects are considered in the simulations, the EOS signal for the 1-mm-thick detection crystal is stronger than for the 500-μm-thick one. However, from

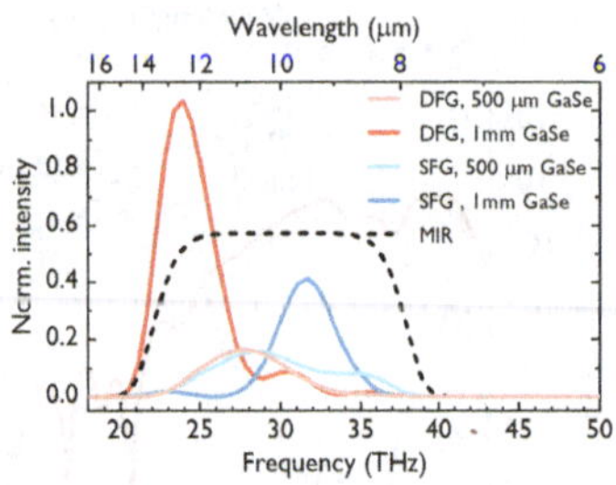

Fig. 3.6 Simulated EOS spectra for SFG and DFG phasematching in a 500-μm and a 1-mm-thick detection crystal for a gate pulse power of 1.5 W. The input MIR spectrum is shown for reference

the experiments, we expect spatial walk-off to play a significant role for the signal strengths (see discussions below).

In conclusion, the simulations suggest that with our gate pulse power, we are at the onset of saturation of the SFG efficiency for thick EOS crystals. With even higher gate pulse powers and thus larger depletion, phasematching DFG instead of SFG can become beneficial. However, the nonlinearity of the signal with the amplified MIR field, especially for higher average powers is to be investigated and the damage threshold of GaSe sets an upper limit to the gate pulse peak intensity. Furthermore, at the gate pulse wavelength of 1960 nm, the currently only available choice of detectors for the DFG signal is extended InGaAs diodes. The increased noise equivalent power for those detectors as compared to standard InGaAs diodes has to be taken into account for the experimental realization.

3.3 Temporal and Spectral Transfer Functions

Due to phasematching, the finite duration of the gate pulse and transmission or reflection properties of the crystal, transport optics and spectral filters, the measured EOS trace is not identical to the incident field. In [6], the distortion of THz waves upon detection with EOS was illustrated. Furthermore, the complex frequency response of electro-optic sensors was introduced in [7], discussed further in [8, 9] and also applied in [10]. Faure et al. [11] contains a detailed discussion of the spectral response, split into its various contributions. In addition, considering the dependence of the EOS signal on the MIR polarization, Oguchi et al. [12] shows how to retrieve the undistorted THz waveform for an elliptically polarized field. The measurement of the electric field or its Hilbert transform is discussed in [1].

Here, the EOS instrument response is calculated numerically, to allow for characterization of the MIR fields, mostly investigating spectral effects and no polarization or spatial dependencies. For a generic complex MIR spectral amplitude $E_{\mathrm{MIR}}(\Omega)$, we obtain the complex EOS spectral amplitude $S_{\mathrm{EOS}}(\Omega)$ as the Fourier transform of the time-domain trace $S_{\mathrm{EOS}}(\tau)$, calculated with the theoretical model described in Sect. 2.3. The variable input parameters to the simulations are the following:

- Peak intensity, as well as spectrum and spectral phase of the gate pulse, typically characterized with FROG [2] for the experimental parameters.
- Thickness and phasematching angles θ and φ of the GaSe detection crystal. The angles are calculated as internal angles with SNLO [13] for phasematching of specific wavelengths, or measured external rotation angles for the experimental configurations.
- Transmission properties of the spectral filters used for dynamic range optimization [14].

The complex frequency-domain IRF as a function of the MIR frequency Ω is then calculated as the ratio of the two complex spectral amplitudes:

$$\mathrm{IRF}(\Omega) = \frac{S_{\mathrm{EOS}}(\Omega)}{E_{\mathrm{MIR}}(\Omega)} \tag{3.1}$$

The time-domain instrument response is obtained from the spectral response *via* the inverse Fourier transform:

$$\mathrm{IRF}(t) = \mathcal{F}^{-1}\{\mathrm{IRF}(\Omega)\} \tag{3.2}$$

This time-domain response can be interpreted as the effective interaction time window at each delay and therefore the time-domain function with which the investigated field was convoluted to obtain the measured trace. For thick detection crystals in particular, this can be much longer than the initial gate pulse duration.

For a known crystal geometry and gate pulse, the measured trace can be corrected with the calculated spectral IRF, to retrieve an estimation of the electric-field waveform. Most importantly for our applications, this back-calculation shows the effects of dispersion of the detection crystal, of the gate pulse temporal profile and of the spectral filters used for dynamic range improvement. The calculation allows to disentangle the various contributions on the response. However, the carrier-envelope phase remains undetermined, because e.g., not all transport optics are considered in the calculations [1] and only the approximate detection crystal thickness is known. In the following experimental chapters, the waveform is typically retrieved for thin detection crystals, for which spatial effects are less severe and the spectral response is more broadband and less modulated, avoiding division by small numbers that can cause instabilities. For validation of the retrieval, the MIR spectrum can be measured with an FTIR spectrometer. To determine the accuracy of the calculated instrument response, e.g., a variation of the phasematching angle in experiment and theory could be compared. If necessary for the investigation of field-dependent effects, the electric field could be measured more precisely using a thin detection crystal as presented in [1], which would allow for the experimental determination of the response function for thicker crystals.

For the field-resolved spectroscopic measurements in our working group, the exact knowledge of the instrument response is not necessary, as long as it depends linearly on the measured field [15]. In this case, the molecular response can be determined by

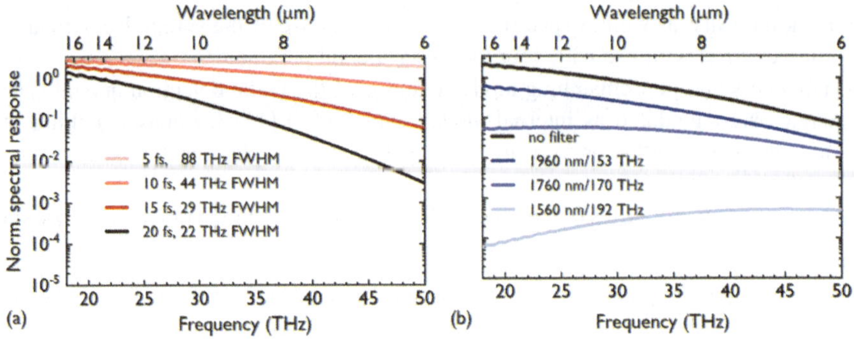

Fig. 3.7 Effect of the gate pulse duration and SPF wavelength on the spectral IRF. **a** Spectral intensity IRF for gate pulse durations varying between 5 and 20 fs. **b** Spectral intensity IRF for varied cut-off wavelength of the SPF used for SNR improvement

comparison of a sample and reference measurement without the field-information. We experimentally verify the linearity by recording EOS traces for various MIR powers (see Sect. 4.9). Especially for thin detection crystals with low MIR depletion, this assumption is justified.

The focus of this thesis was the optimization of the detection efficiency and dynamic range without the exact knowledge of the investigated field. However, two effects affecting the spectral response are discussed here. The requirement of spectral overlap of the local oscillator and generated sum-frequency intrinsically limits the maximum pulse duration or minimum bandwidth of the gate pulse to detect frequencies which cause a large spectral shift of the sum- and difference-frequency fields. To illustrate the effect of the gate pulse duration on the spectral response of the EOS detection, the spectral IRF was simulated for Fourier-transform limited, Gaussian gate pulses at a central wavelength of 1960 nm with FWHM durations varying between 5 and 20 fs and a 1-μm-thick detection crystal. Figure 3.7a shows the resulting responses in the wavelength range from 6 to 16 μm, as this is the experimentally investigated spectral region.

With increasing gate pulse duration, the spectral response drops towards higher frequencies. This is expected from the simple time-domain picture, where the gate pulse has to be shorter than one half-cycle of the measured field in order to temporally resolve the oscillation. In the frequency domain, the decreased response stems from a lack of local oscillator photons, as the high MIR frequencies cause a large shift in the sum-frequency wavelength.

Furthermore, to illustrate the effect of the spectral filter, which can be employed for dynamic range optimization (see Sect. 2.6), the spectral IRF for different filter wavelengths is shown in Fig. 3.7b. Filtering at shorter wavelengths decreases the contributions from low frequencies, because only the highest MIR frequencies mix to the wings of the gate pulse spectrum (see also the illustrations in [1]). However, as discussed in Sect. 2.6 and in the experimental chapters, depending on the spectral shape and the noise properties of the local oscillator, this can still increase the detection dynamic range.

3.4 Detection Bandwidths for Different Gate Pulse Wavelengths

The benefits of using longer-wavelength driving pulses for MIR generation *via* IPDFG have been discussed previously [16, 17]. However, using gate pulses with different central wavelengths for EOS has not been investigated before.

In GaSe—our EOS crystal of choice for broadband MIR detection [18]—moving to longer wavelength gate pulses reduces dispersion and thus improves phase matching conditions. Improved phasematching allows for the use of thicker detection crystals, which is beneficial in several regards: It increases the MIR to NIR conversion efficiency while maintaining a broad spectral coverage and increases the time window without multiple reflections—which will be discussed in detail in Sect. 4.4. Furthermore, using longer gate pulse wavelengths increases the crystal damage threshold for multi-photon absorption and allows for the use of higher gate pulse powers, thus also benefiting the MIR photon upconversion efficiency.

Erbium fiber lasers are commonly used for the generation and field-resolved detection of broadband infrared waveforms [19]. However, the average powers of the gate pulses are limited to several hundred mW. In recent years, high-power laser-frontends with central wavelengths in the infrared were developed. These are based on ytterbium [20–23], thulium [24–26], holmium [27, 28] and chromium technology [29, 30].

Within the scope of this thesis, gate pulses from three different laser systems were investigated experimentally: 1030-nm pulses with 16 fs pulse duration and 450 mW average power, as generated by a 28-MHz, Yb:YAG thin-disk-laser based system [31], 1550-nm pulses with 11 fs pulse duration and 120 mW average power, from a 56-MHz, Er-fiber-laser [32] and 1960-nm pulses with 13 fs pulse duration and 1.9 W average power, as generated by a 50-MHz, Thulium-fiber-based laser system [33].

In the following, the effect of the increasing gate pulse wavelength on the detection bandwidth of EOS is investigated theoretically [34]. To this end, the IRF was calculated for the different gate pulse wavelengths and detection crystal thicknesses varying from 1 to 300 μm. For the 1030 and 1960-nm pulses, this comparison was shown in [34]. First, to separate the effect of the increasing gate pulse wavelength from response changes caused by varying pulse durations, only the gate pulse central wavelength is varied for an otherwise identical, Fourier-transform limited Gaussian pulse shape with 10 fs FWHM pulse duration. For each wavelength, the GaSe rotation angle was chosen to phasematch the central wavelength of the gate pulse and the 10-μm field. This corresponds to $\theta_{1030} = 16.1°$, $\theta_{1550} = 13.5°$ and $\theta_{1960} = 12.7°$, as calculated with SNLO [13]. The resulting frequency-domain intensity responses are shown in Fig. 3.8.

For the gate pulse with 1030 nm central wavelength, the IRF has a maximum around 10 μm for the thinnest crystals. It narrows with increasing crystal thickness and the maximum shifts to shorter wavelengths. In the case of the 1550 nm gate pulses, the maximum of the spectral response is at the longest MIR wavelength that is considered here and stronger than for the shorter gate pulse wavelength. The

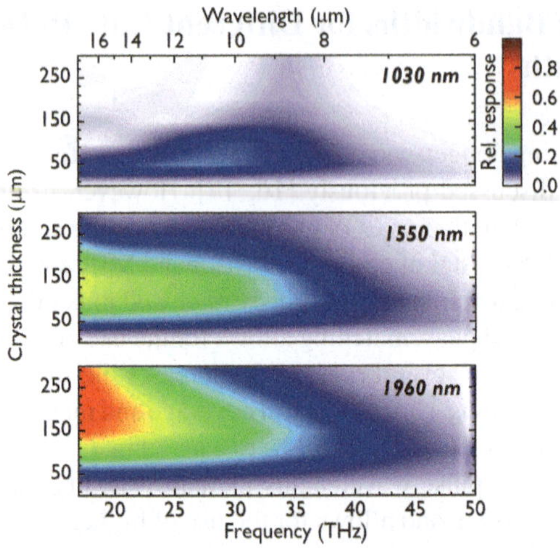

Fig. 3.8 Simulated frequency-domain IRF for gate pulses with a Gaussian intensity envelope with 10 fs FWHM pulse duration and for varying detection crystal thicknesses. The central wavelength of the gate pulse is 1030 nm in the top panel, 1550 nm in the middle and 1960 nm on the bottom. For each wavelength, a spectral filter transmitting only wavelengths below the central wavelength was used in the simulation of the heterodyne detection

IRF becomes more narrowband with increasing crystal thickness. A similar effect is observed for the longest gate pulse wavelength of 1960 nm. In this case, the maximum of the response is even higher and the effect of spectral narrowing is less severe. All gate pulse wavelengths show a drop of the spectral response towards shorter MIR wavelengths. This trend can be explained by the effect of the gate pulse duration as discussed before, see Fig. 3.7a. The monotonous dispersion of GaSe leads to improved phasematching for longer-wavelength gate pulses, which results in the broadening of the IRF from top to bottom. Furthermore, an increase in the strength of the spectral response and a shift of the maximum to thicker detection crystals is observed when increasing the gate pulse wavelength. These effects can be explained by the reduced dispersion of GaSe for longer wavelengths: The GDD decreases from 1030 fs^2/mm at 1030 nm to 408 fs^2/mm at 1960 nm. This reduction of the dispersion by more than a factor of two causes the gate pulse to disperse less in time. Thus, the average peak intensity of the gate pulse inside the detection crystal, which directly affects the detection efficiency, is higher.

Next, the IRF was compared for the gate pulses that were used for the experimental results shown in the next chapter. The electric field of the gate pulses was retrieved from FROG measurements of the temporally compressed experimental pulses (see Fig. 4.2 for 1030 nm, Fig. 4.19 for 1550 nm and Fig. 4.29 for 1960 nm). The internal angle of the beams inside the GaSe crystal was set for phasematching 8 μm in the MIR and the central wavelength of the gate pulse ($\theta_{1030\,\mathrm{nm}} = 18°$, $\theta_{1550\mathrm{nm}} = 14.8°$, $\theta_{1960\,\mathrm{nm}} = 13.7°$, $\varphi = 0°$ for all cases). Spectral filters for dynamic range improvement (see Sect. 2.6 were implemented with cut-off wavelengths similar to

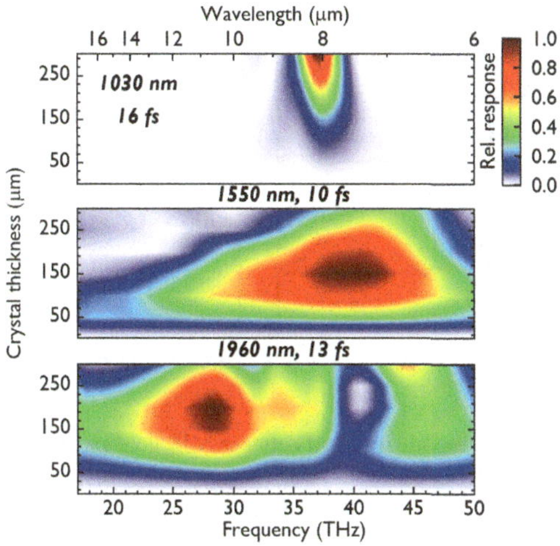

Fig. 3.9 Comparison of simulated frequency-domain IRF for experimentally available gate pulses and varying detection crystal thicknesses. Top: 16-fs-long gate pulses at 1030 nm central wavelength, 912 nm SPF. Middle: 11-fs-long gate pulses at 1550 nm central wavelength, 1400 nm SPF. Bottom: 13-fs-long gate pulses at 1960 nm central wavelength, 1600 nm SPF. All IRFs are normalized to the maximum for each gate pulse wavelength

what was used in the experimental setups described below: A 912-nm SPF for the 1030-nm gate pulses, a 1400-nm SPF for the 1550-nm gate pulses and a 1600-nm SPF for the 1960-nm gate pulses. The frequency-domain IRFs for the investigated gate pulses are compared in Fig. 3.9.

For 1030-nm gate pulses (top panel), the IRF has a sharp maximum at 8 μm, that decreases in width and shifts slightly to the red with increasing crystal thickness. Even the thinnest investigated GaSe crystals do not support a spectrally flat response with this gate pulse wavelength. With the 1550-nm gate pulses a much broader spectral coverage is reached. For the 1-μm-thick crystal, the spectral response covers the full investigated spectral range from 6 μm to 16 μm with a minimum in the relative response varying between 60 and 100 %. The FWHM of the spectral response gradually narrows with the increasing detection crystal thickness, spanning from 6.4 to 8.2 μm for the 300-μm-thick one. The peak of the spectral response shifts to higher MIR frequencies with increasing crystal thickness. This behavior is most likely caused by decreased dispersion of GaSe at lower gate pulse frequencies, which disperse less in thicker crystals and therefore mix more efficiently. Using a SPF for the local oscillator requires low-frequency gate photons to most efficiently detect high-frequency MIR photons [1]. For the 1960-nm gate pulses, the IRF has a 'double-humped' appearance, covering the spectral range up to 15 μm at half the response maximum. It narrows to a FWHM detection bandwidth spanning from 6.7 to 11.7 μm and the maximum response is also shifted towards shorter wavelengths. The modulation of the spectral response can be explained by the two-lobe structure of the gate spectrum, which drops almost to zero intensity around the center

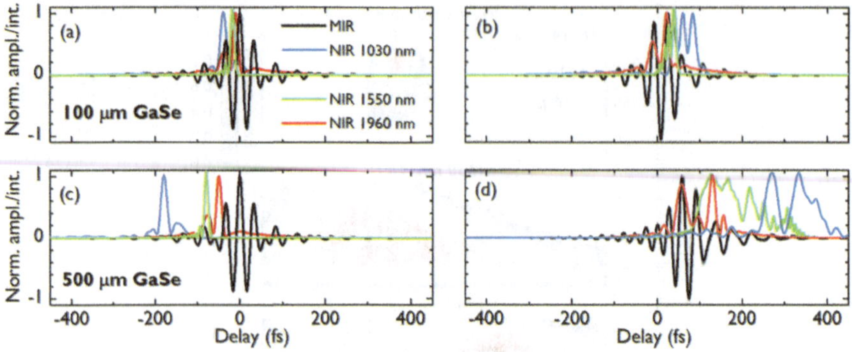

Fig. 3.10 Comparison of the temporal walk-off between MIR field and gate pulse temporal intensity envelope for different gate pulses and crystal thicknesses. In all cases, the delay point of maximum EOS signal is illustrated. **a**, **b** MIR field and gate pulse intensity envelopes at the beginning and end of a 100-μm-thick GaSe crystal, respectively. **c**, **d** MIR field and gate pulse intensity envelopes at the beginning and end of a 500-μm-thick GaSe crystal, respectively

wavelength. Comparing the maxima of the spectral response as a function of the crystal thickness shows that the detection efficiency increases for the 16-fs pulses at 1030 nm, while a maximum is reached for lower thicknesses for the other two gate pulses. One explanation for this effect is the increasing effect of dispersion on the shorter gate pulses.

In the time-domain, the limited phase matching bandwidth can be understood by the temporal walk-off of the interacting pulses, caused by the group-velocity mismatch. For the monotonously changing dispersion of GaSe, the temporal walk-off reduces for MIR and gate pulse wavelengths lying closer together, i.e., for moving the gate pulse wavelengths further towards the infrared. Furthermore, as it was already mentioned for the comparison of the phasematching bandwidths, the dispersion of the gate pulses is reduced at longer central wavelengths. To illustrate these effects for the increasing gate pulse wavelengths, the infrared fields and gate pulse temporal intensity envelopes were plotted at the beginning and at the end of the EOS crystal at the delay position of maximum signal, see Fig. 3.10.

Figure 3.10a shows the start conditions for the gate pulse and infrared field entering the crystal for maximum EOS signal, while the temporal intensity profiles of the gate pulses after the 100-μm-thick crystal and their timing with respect to the investigated waveform is shown in Fig. 3.10b. Passing through the GaSe crystal chirps the initially compressed gate pulses in time. The FWHM pulse duration of the pulses with 1030 nm central wavelength increases from 16 to 60 fs. This effect is reduced at longer gate pulse wavelengths, i.e., at 1960 nm, the 13.8 fs pulses mostly show an increase of the side-pulses without an increase of the main-pulse duration. Figure 3.10c, d show the start and end conditions for optimum signal strength for a 500-μm-thick crystal. This increases the dispersion, causing all gate pulses to extend over the full MIR field at the end of the crystal and a chirp of the initially compressed MIR waveform. In addition to the temporal dispersion, a decrease of the temporal

Table 3.1 Relative temporal walk-off between the peak of the MIR field and the center of mass of the gate pulse temporal intensity envelope for gate pulses at different central wavelengths and two different detection crystal thicknesses

GaSe thickness (μm)	1030 nm gate pulse (fs)	1550 nm gate pulse (fs)	1960 nm gate pulse (fs)
100	98	52	28
500	489	261	142

walk-off with increasing gate pulse wavelength can be observed. The calculated temporal walk-off at different gate pulse wavelengths in 100 and 500 μm GaSe crystals is summarized in Table 3.1.

Gradually increasing the gate-pulse wavelength from 1030 to 1960 nm decreases the temporal walk-off from 98 to 28 fs in a 100-μm-thick GaSe crystal. A reduction by a similar factor is observed in the five times thicker crystal. The group velocities of the MIR and gate pulse are more similar for longer gate pulse wavelengths, thus causing the smaller walk-off. This reduced temporal walk-off directly relates to the increase in the phasematching bandwidth observed in Figs. 3.8 and 3.9.

In conclusion, several effects of the reduced dispersion of GaSe for longer wavelengths on the EOS detection were illustrated with simulation results for different gate pulse wavelengths. These effects are a decrease of the GDD and thus a reduced temporal stretching of the gate pulses and a reduction in the temporal walk-off between MIR field and gate pulse, resulting in an increase in the phasematching bandwidth. In Chap. 4, the benefits of choosing gate pulses with longer central wavelengths will be investigated experimentally. Increasing the gate-pulse center wavelength to beyond 1960 nm has become possible with the temporal compression of the output pulses of chromium laser frontends [29, 30] and is currently investigated in our working group. An upper limit to the gate-pulse wavelength is reached when it spectrally overlaps with the characterized radiation, because this would result in a constant background signal.

3.5 Pre-chirping of 1550-nm Gate Pulses

As illustrated in the previous section, the dispersion of GaSe causes an initially compressed gate pulse to disperse significantly in time and thus limits the signal strength achievable with short gate pulses in thick detection crystals. Pre-chirping the gate pulse before the detection crystal increases the average peak intensity in the nonlinear medium and flattens the phase relation (see Sect. 3.1) between the gate and generated sum-frequency field, and thus increases the signal strength. While a detailed investigation of pre-chirping would benefit the signal strength in all EOS setups discussed below, it was experimentally studied in detail for the 1550-nm gate pulses described in Sect. 4.6. Here, the effect is investigated theoretically for

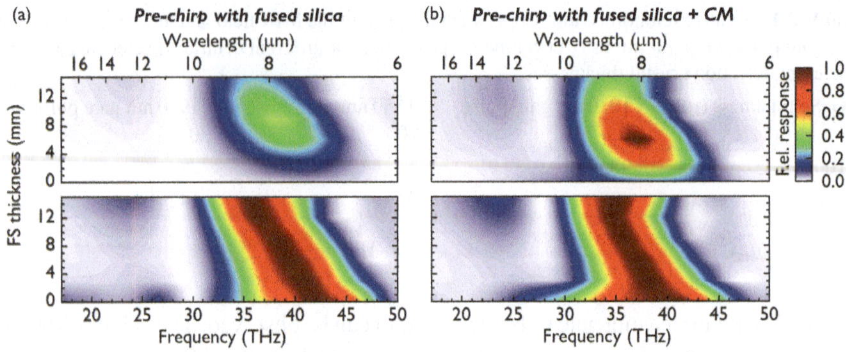

Fig. 3.11 Simulated EOS spectral intensity IRF for several methods of pre-chirping the 1550-nm gate pulses. **a** Pre-chirp with fused silica. Top: Quantitative comparison of the non-normalized spectral intensity IRF. Bottom: IRF normalized to the maximum for each fused silica thickness for comparison of the spectral coverage. **b** Pre-chirp with fused silica and chirped mirrors (CM). Top: Non-normalized spectral intensity IRF, allowing for comparison of the signal strenghts. Bottom: IRF normalized to the maximum for each fused silica thickness for comparison of the spectral coverage

deeper understanding. The 1550-nm gate pulses are shown in Fig. 4.19 and a 300-μm-thick crystal was used for EOS detection. The spectral phase of fused silica, with thicknesses varied between 0 and 15 mm, was added to the gate pulses before calculating the instrument response. Fused silica is a suitable pre-chirping material for wavelengths above 1270 nm, where its GDD has the opposite sign as compared to that of GaSe. The angle of the beams inside the GaSe crystal was set to $\theta = 14.8°$ and $\varphi = 0°$, for phasematching SFG of 8 μm and 1550 nm, propagating on the extraordinary and ordinary crystal axis, respectively. Furthermore, a SPF at 1400 nm was implemented, matching that used in the experiment. Figure 3.11a shows the frequency-domain IRF for varied fused silica thickness.

A quantitative comparison of the strength of the spectral response for different pre-chirps is possible in the top panel, while the responses in the lower panel were normalized to the maximum for each thickness of fused silica to illustrate its spectral evolution. The maximum response is achieved for pre-chirping with a 9-mm-thick fused silica plate. This amount of fused silica has the negative GDD of half of the detection crystal thickness (150 μm GaSe) at ≈ 1400 nm. With increasing fused silica thickness, the maximum of the spectral response shifts from 7.1 to 8.3 μm. Both the maximum and the spectral shift can be explained by the dispersion properties of fused silica. Increasing the amount of fused silica optimizes the pre-chirping for shorter gate pulse wavelengths, thus favoring the detection of longer wavelength MIR radiation (the opposite effect as described for Fig. 3.9). Most likely, the strongest wavelength dependent electro-optic signal occurs around 1400 nm, resulting in the optimum for best compression at this wavelength. In future experiments, pre-chirping will be further optimized by using chirped mirrors, which compensate for the GDD of 150 μm of GaSe (half the EOS crystal thickness) over the full spectral bandwidth of the gate pulses. Figure 3.11b shows the expected instrument response with the designed phase of the chirped mirrors applied to the gate pulses, again for different amounts of fused silica in the gate pulse beam path. While the overall trend of the spectral response is comparable to panel (a), two main differences are observable: For the same phasematching angle, the spectral response is shifted to longer wavelengths, with the maximum now varying between 7.6 and 8.3 μm with increasing fused silica thickness. Furthermore, the maximum response is more than a factor of two stronger and is now achieved for adding 5 mm of fused silica. The chirped mirrors compress the gate pulses in the centre of the crystal and thus maximize the sum-frequency generation efficiency. In addition, adding fused silica can still improve the phase relations in the wavelength dependent signal and thus increase the signal strength. The effect of the chirped mirrors has in the meanwhile also been observed experimentally.[1]

In conclusion, pre-chirping 1550 nm gate pulses with fused silica increases the EOS signal strength. However, dispersion compensation can only be achieved for wavelengths longer than 1270 nm, where the GDD of GaSe and fused silica have opposite signs. Due to the different shapes of the dispersion curves of GaSe and fused silica, perfect pre-chirping can also only be achieved at one wavelength. The effect of pre-chirping is improved by the use of chirped mirrors that match the spectral phase of GaSe over a broader spectral range.

References

1. Sulzer P et al (2020) Determination of the electric field and its Hilbert transform in femtosecond electro-optic sampling. Phys Rev A 101:033821. ISSN: 1050-2947
2. Trebino R et al (1997) Measuring ultrashort laser pulses in the time-frequency domain using frequency-resolved optical gating. Rev Sci Instrum 68:3277–3295. ISSN: 0034-6748
3. Beckh C, Sulzer P, Fritzsche N, Riek C, Leitenstorfer A (2021) Analysis of subcycle electro-optic sampling without background. J Infrared Millimet Terahertz Waves 350:420. ISSN: 1866-6892
4. Cornet M, Degert J, Abraham E, Freysz E (2016) Frequency resolved cross-correlation between optical and terahertz pulses: application to ultrashort laser pulse characterization. Opt Exp 24:3003–3010. ISSN: 1094-4087
5. Boyd RW (2008) Nonlinear optics, 3rd edn. Academic Press, Burlington
6. Bakker HJ, Cho GC, Kurz H, Wu Q, Zhang X-C (1998) Distortion of terahertz pulses in electro-optic sampling. J Opt Soc Am B 15:1795. ISSN: 0740-3224
7. Leitenstorfer A, Hunsche S, Shah J, Nuss MC, Knox WH (1999) Detectors and sources for ultrabroadband electro-optic sampling: experiment and theory. Appl Phys Lett 74:1516–1518. ISSN: 0003-6951
8. Yellampalle B, Kim KY, Rodriguez G, Glownia JH, Taylor AJ (2005) Algorithm for high-resolution single-shot THz measurement using in-line spectral interferometry with chirped pulses. Appl Phys Lett 87:211109. ISSN: 0003-6951

[1] Chirped mirror coating by V. Pervak (Coating number CM2051, experimental implementation by P. Jacob and A. Weigel.

9. Jamison SP, MacLeod AM, Berden G, Jaroszynski DA, Gillespie WA (2006) Temporally resolved electro-optic effect. Opt Lett 31:1753–1755. ISSN: 1539-4794
10. Keiber S et al (2016) Electro-optic sampling of near-infrared waveforms. Nat Photon 10:159–162. ISSN: 1749-4885
11. Faure J, van Tilborg J, Kaindl RA, Leemans WP (2004) Modelling laser-based table-top THz sources: optical rectification, propagation and electro-optic sampling. Opt Quant Electron 36:681–697. ISSN: 0306-8919
12. Oguchi K, Okano M, Watanabe S (2017) Retrieving the undistorted terahertz timedomain electric-field vector from the electro-optic effect. J Opt Soc Am B 34:1946. ISSN: 0740-3224
13. SNLO nonlinear optics code available from A. V. Smith, AS-Photonics, Albuquerque, NM
14. Porer M, Ménard J-M, Huber R (2014) Shot noise reduced terahertz detection via spectrally postfiltered electro-optic sampling. Opt Lett 39:2435–2438. ISSN: 1539-4794
15. Pupeza I et al (2020) Field-resolved infrared spectroscopy of biological systems. Nature 577:52–59. ISSN: 1476-4687
16. Junginger F et al (2010) Single-cycle multiterahertz transients with peak fields above 10 MV/cm. Opt Lett 35:2645–2647. ISSN: 1539-4794
17. Petrov V (2015) Frequency down-conversion of solid-state laser sources to the mid-infrared spectral range using non-oxide nonlinear crystals. Progr Quant Electron 42:1–106. ISSN: 00796727
18. Kübler C, Huber R, Tübel S, Leitenstorfer A (2004) Ultrabroadband detection of multi-terahertz field transients with GaSe electro-optic sensors: approaching the near infrared. Appl Phys Lett 85:3360–3362. ISSN: 0003-6951
19. Sell A, Leitenstorfer A, Huber R (2008) Phase-locked generation and field-resolved detection of widely tunable terahertz pulses with amplitudes exceeding 100 MV/cm. Opt Lett 33:2767. ISSN: 0146-9592
20. Brons J et al (2016) Powerful 100-fs-scale Kerr-lens mode-locked thin-disk oscillator. Opt Lett 41:3567–3570. ISSN: 1539-4794
21. Saltarelli F et al (2019) Power scaling of ultrafast oscillators: 350-W average-power subpicosecond thin-disk laser. Opt Exp 27:31465–31474
22. Saraceno CJ, Sutter D, Metzger T, Ahmed M (2019) The amazing progress of high-power ultrafast thin-disk lasers. J Eur Opt Soc Rapid Publ 15:1–7. ISSN: 1990-2573
23. Müller M et al (2020) 10.4 kW coherently combined ultrafast fiber laser. Opt Lett 45:3083–3086. ISSN: 1539-4794
24. Wan P, Yang L-M, Liu J (2013) High power 2 μm femtosecond fiber laser. Opt Exp 21:21374–21379
25. Stutzki F et al (2014) 152 W average power Tm-doped fiber CPA system. Opt Lett 39:4671–4674. ISSN: 1539-4794
26. Zhang J et al (2018) High-power, high-efficiency Tm:YAG and Ho:YAG thin-disk lasers. Laser Photon Rev 12:1700273. ISSN: 18638880
27. Li P, Ruehl A, Grosse-Wortmann U, Hartl I (2014) Sub-100 fs passively mode-locked holmium-doped fiber oscillator operating at 2.06 μm. Opt Lett 39:6859–6862. ISSN: 1539-4794
28. Zhang J, Mak KF, Pronin O (2018) Kerr-Lens mode-locked 2-μm thin-disk lasers. IEEE J Select Top Quant Electron 24:1–11. ISSN: 1558-4542
29. Nagl N et al (2019) Directly diode-pumped, Kerr-lens mode-locked, few-cycle Cr:ZnSe oscillator. Opt Exp 27:24445–24454
30. Vasilyev S et al (2019) Super-octave longwave mid-infrared coherent transients produced by optical rectification of few-cycle 25-μm pulses. Optica 6:111. ISSN: 2334-2536
31. Pupeza I et al (2015) High-power sub-two-cycle mid-infrared pulses at 100 MHz repetition rate. Nat Photon 9:721–724. ISSN: 1749-4885
32. Brida D, Krauss G, Sell A, Leitenstorfer A (2014) Ultrabroadband Er: fiber lasers. Laser Photon Rev 8:409–428. ISSN: 18638880

33. Butler TP et al (2019) Watt-scale 50-MHz source of single-cycle waveform-stable pulses in the molecular fingerprint region. Opt Lett 44:1730–1733. ISSN: 1539-4794
34. Hofer C et al (2019) Quantum-efficiency and bandwidth optimized electro-optic sampling. In: 2019 conference on lasers and electro-optics Europe & European quantum electronics conference (CLEO/Europe-EQEC). IEEE, Piscataway

Chapter 4
Experimental Results

In the previous chapter, a theoretical analysis of the phasematching and dispersion advantages of using gate pulses with longer central wavelengths was presented. Ultimately, increasing the gate pulse wavelength and thus being able to use thick detection crystals, while maintaining octave spanning detection bandwidths, has the goal of increasing the MIR photon detection efficiency. As compared to linear characterization techniques, which rely on low-noise detectors for a high detection sensitivity, the nonlinear effect employed for EOS detection intrinsically causes photon losses in the upconversion step. Furthermore, in the second step of heterodyne detection, an imperfect spatial overlap and spectral filtering reduce the number of detected photons. Overall, we thus define the detection efficiency as the ratio of the number of sum-frequency photons contributing to the electro-optic signal, to the number of MIR photons impinging on the detection crystal. So far, the estimation presented in [1], with an approximate upconversion efficiency of 0.1%, is the only reference for the efficiency of the electro-optic detection.

Here, high-power gate pulses and a detailed analysis of photon losses in the two-step process are employed to characterize and optimize the number of detected MIR photons for the first time. Measuring the depletion of the MIR power as a function of the gate pulse delay is employed for investigating the upconversion efficiency. The first optimization of the detection efficiency was performed with 1-μm gate pulses, resulting in $\approx 0.76\%$ maximum overall detection efficiency, and an unprecedented measurement dynamic range in the MIR spectral range that was applied for the spectroscopic measurements presented in [2]. For these gate pulses, the trade-off between detection efficiency and bandwidth is illustrated next, followed by a discussion of several approaches to avoid detecting multiple reflections from the GaSe detection crystal.

© The Author(s), under exclusive license to Springer Nature Switzerland AG 2022
C. Hofer, *Detection Efficiency and Bandwidth Optimized Electro-Optic Sampling of Mid-Infrared Waves*, Springer Theses,
https://doi.org/10.1007/978-3-031-15328-0_4

One application of the high detection dynamic range is the measurement of the stability of the MIR waveforms and the theoretical analysis of the correspondence between their measured fluctuations and those of the electric field. Furthermore, EOS detection with 1550 nm gate pulses is compared to gating at 1030 nm, elucidating the improved phasematching for longer gate pulse wavelengths. Finally results for EOS with 2-μm gate pulses are presented. The high gate pulse power and even better phasematching than at 1550 nm, allowed us to reach multi-percent-level detection efficiencies. For these high detection efficiencies, the linearity of the signal with the infrared field strength and gate pulse power is investigated.

4.1 Optimization and Characterization of EOS with 1-μm Gate Pulses

The first optimization and characterization of the EOS detection efficiency was performed with 1030-nm gate pulses from a high-power Yb:YAG thin-disk laser system. Figure 4.1 shows the schematic of the experimental setup,[1] which is described in detail in [3].

An Yb:YAG thin-disk oscillator[2] [4], delivers 220-fs (FWHM pulse duration) pulses centered at 1030 nm wavelength with 90 W of average power at 28 MHz repetition rate. The pulses are spectrally broadened in three Herriott-cells [4, 5] and temporally compressed with dispersive mirrors. After the third cell, a small portion of the power is transmitted through a 920-nm SPF to generate an error signal for power stabilization with an acousto-optic modulator (AOM) at the oscillator output [6]. The remaining 16-fs pulses with 60 W of average power are sent to a 1-mm-thick lithium gallium sulfide ($LiGaS_2$, LGS) crystal for IPDFG [7], resulting in infrared fields with a maximum of 60 mW average power and centered at 8 μm wavelength. The generated long-wavelength radiation and the NIR driving pulses are separated with a dichroically coated ZnSe plate. A mechanical chopper modulates the MIR beam at 7 kHz for lock-in detection of the EOS signal. The 8-μm pulses are then transmitted through a sample cell and can be temporally compressed with chirped mirrors [8]. A wedge reflection of the NIR pulses after the IPDFG crystal is temporally re-compressed with chirped mirrors, delayed with a mechanical stage and variably attenuated with a pellicle on a rotation mount. A two-lens telescope adjusts the focus size at the position of the EOS crystal to that of the MIR beam. Placing the second lens (L2) on a translation stage allows for collimation adjustment and thus setting of the lateral focus position. The MIR and gate beams are spatially combined with an uncoated germanium plate, placed at Brewster's angle for minimum losses. An auxiliary laser tracks the relative interferometer length between the two channels

[1] K. Fritsch and S. A. Hussain built the laser frontend with the help of W. Schweinberger and M. Huber. I planned, built, optimized and characterized the EOS detection (indicated by the green dashed line in Fig. 4.1), with the help of S. A. Hussian and W. Schweinberger.

[2] developed in the group of O. Pronin by K. Fritsch

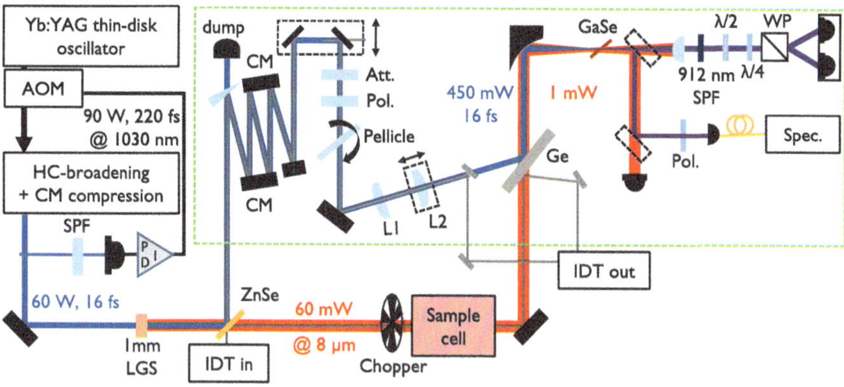

Fig. 4.1 Schematic of the laser system used for the optimization and characterization of the detection efficiency of EOS with 1-μm gate pulses. An Yb:YAG thin-disk oscillator drives three Herriott-cell-type spectral broadening and chirped mirror compression stages. With an acousto-optic modulator (AOM) placed after the oscillator, the power measured behind a SPF at 920 nm is stabilized after the last compression stage. With the resulting 1030-nm pulses with 60 W average power and 16 fs pulse duration, radiation with an average power of 60 mW and with a central wavelength of 8 μm is generated *via* IPDFG in a 1-mm-thick LGS crystal. The driving pulses are then split-off with a dichroically coated ZnSe plate. A fraction of these pulses is re-compressed with chirped mirrors (CM), attenuated and polarization cleaned to serve as the EOS gate pulse. A rotatable pellicle allows for variable power attenuation. The gate pulse is variably delayed with a mechanical stage and a lens-telescope decreases its beam size. The MIR beam is mechanically chopped and transmitted through a sample cell for biological measurements, before it is spatially combined with the gate beam using a germanium plate (Ge). The relative interferometer length is tracked with an auxiliary laser (interferometric delay tracking (IDT)). An off-axis parabolic mirror focuses the recombined beams into a GaSe crystal for EOS detection. Using a flip-mirror, the MIR and sum-frequency beams can be characterized. For the signal measurement, the gate and sum-frequency beams are transmitted through a 912-nm SPF. Two waveplates and a Wollaston prism (WP) serve as the ellipsometry setup before balanced detection

(interferometric delay tracking (IDT) [9]). For EOS detection, the spatially combined beams are focused into a GaSe crystal using an off-axis parabolic mirror with 50 mm effective focal length. After the nonlinear crystal, the MIR photon upconversion efficiency is measured by tracking the transmitted power of the characterized pulses, using a μW power sensor from LASNIX. Additionally, the generated sum-frequency spectrum is measured after transmission of the NIR beam through a polarizer for suppression of the remaining gate pulse power. A lens collimates the generated sum-frequency and the local oscillator beam and a 912-nm SPF is used for dynamic range optimization [1], see calculation in Sect. 2.6. A combination of two achromatic waveplates is used to balance the amount of power sent to the two diodes after beam-splitting with a Wollaston prism (Thorlabs WP10). The electro-optic signal is detected with a home-built balanced detector, using two diodes (First Sensor, PC5-7 TO) with a quantum efficiency of 100% at 900 nm wavelength. The differential output current of the balanced detector is amplified with a current amplifier (DLPCA-200 from FEMTO) and then detected with a lock-in amplifier (Zurich Instruments,

MFL) at the modulation frequency of the chopper placed in the MIR beam path. The electro-optic signal is not normalized to the total power on the balanced diodes. While this would reduce fluctuations of the signal strength caused by the gate-pulse power, it would introduce nonlinearities when significant sum-frequency powers are generated.

As derived in Sect. 2.3, the signal strength scales with the peak intensity of the gate pulse and the amplitude of the MIR field. Therefore, the first step of the EOS signal strength optimization was the temporal recompression of the gate pulses after dispersion in the IPDFG crystal using dispersive mirrors and accounting for the additional material in the gate pulse beam path. The temporal intensity profile at the detection crystal was characterized using a home-built FROG [10] setup with a 10-μm-thick BBO crystal, with the results shown in Fig. 4.2.

Panel (a) shows the retrieved spectrum and spectral phase and an independently measured reference spectrum. The measured and retrieved spectra agree up to a small shift in central wavelength, validating the FROG measurement. Third-order contributions to the spectral phase cause the side pulses that are visible in the retrieved temporal intensity envelope in panel (b). Reduction of the side pulses would require the use of chirped mirrors with a tailored spectral phase.

In addition to the gate pulse (peak) intensity, the SFG efficiency depends on the spatial overlap of the MIR and NIR beams, as described by Eq. (2.37) [11]. A two-lens telescope in the gate pulse beam path (lenses L1 and L2, see Fig. 4.1), reduces its beam diameter, such that the focus sizes of the interacting beams are matched at the position of the GaSe crystal. The lateral position of the NIR focus was set *via* translation of the second lens of the telescope to match that of the MIR focus. The resulting focus spot sizes for both beams were measured with CCD cameras (WinCamD-UHR and WinCamD-FIR2-16-HR, respectively) and are shown in Fig. 4.2c. The $1/e^2$ beam diameters in x and y direction are $2w_x = 70\,\mu m$ and $2w_y = 60\,\mu m$ for the gate beam and $2w_x = 90\,\mu m$ and $2w_y = 75\,\mu m$ for the MIR beam.

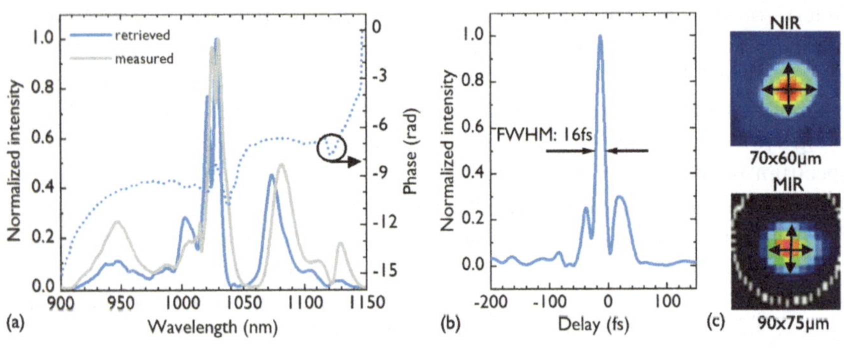

Fig. 4.2 Gate-pulse FROG measurement and beam profiles at the EOS crystal. **a** Retrieved gate pulse spectrum/spectral phase and independently measured spectrum for validation. **b** Retrieved temporal intensity envelope, FWHM pulse duration of 16 fs. **c** MIR and NIR beam profiles at EOS crystal position with $1/e^2$ diameters, axes indicated with arrows

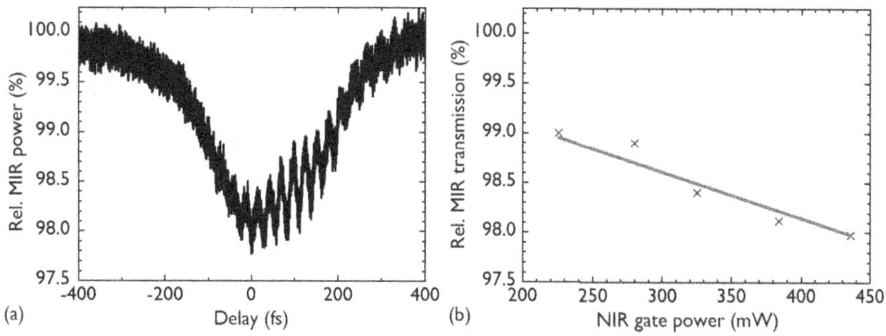

Fig. 4.3 **a** Measurement of the MIR depletion as a function of EOS delay for a 500-μm-thick EOS crystal. **b** Maximum of the delay-dependent depletion for varied gate pulse power. Solid line: linear fit, intersecting the y-axis at 100%, with a slope of $0.00464 \pm 0.00016\,\%/\mathrm{mW}$

For the given NIR focus spot size, we then maximized the gate pulse power. We used a beamsplitter (BS502, custom in-house coating by V. Pervak), polarizer and a 100-nm-thin, silicon nitride (Si_3N_4) pellicle for power tuning. Above 500 mW of gate pulse power reaching the GaSe crystal, it showed flickering as a first sign of damage. For stable operation, we set the gate power to 450 mW. As demonstrated in [1], a SPF was used for optimization of the DR for a given phasematching configuration. When optimizing the detection for 8-μm MIR radiation, with our gate pulses, this is achieved with a 912 nm SPF, matching the prediction by the calculation in Sect. 2.6.

To characterize the efficiency of the MIR to sum-frequency conversion, we measured the MIR depletion behind the electro-optic crystal. A thermal power sensor (μW power sensor for mid-IR laser beams, 511 from LASNIX) tracked the power transmitted through the EOS crystal while the gate pulse delay was scanned over the signal maximum. Figure 4.3a shows the resulting depletion as a function of the gate pulse delay for a 500-μm-thick detection crystal.

At the minimum, the transmitted MIR power drops by 2%. After the minimum, the power measurement shows oscillations with a frequency matching that of the 8-μm field. These oscillations are most likely caused by interference of radiation generated in the EOS crystal with the transmitted MIR beam. This effect was minimized by careful adjustment of the NIR polarizer to avoid DFG in the detection crystal.

In order to relate the measured 2% depletion to the number of photons that can interact with the gate pulse at the delay position of maximum signal, the effective interaction time window has to be considered. For the 500-μm-thick crystal, the temporal walk-off between the MIR and NIR pulses is 489 fs (see Table 3.1). This means that the gate pulse interacts with almost all photons of the 60-fs-long MIR pulse (determined from the retrieved field in Fig. 4.10 and discussed below) at the delay corresponding to maximum EOS signal. From the nonlinear polarization for SFG, it follows that the power depletion of the measured field scales with the gate pulse power. To validate this scaling law for the significant reduction in MIR power, the depletion was measured for several gate pulse powers, varied by rotation of the

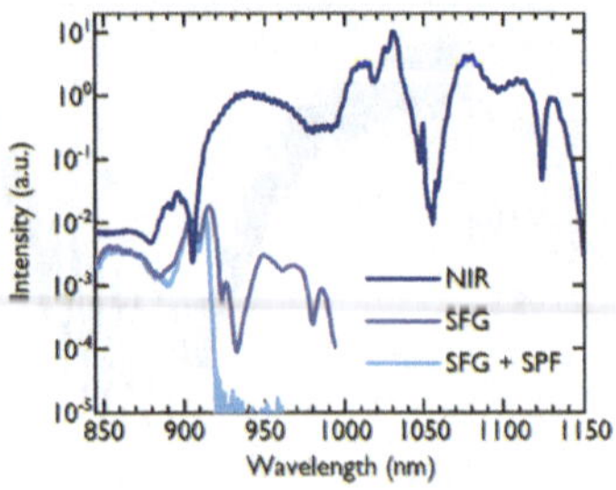

Fig. 4.4 Measurement of gate and sum-frequency spectrum with an optical spectrum analyzer, showing the spectral cut-off of the SPF used for dynamic range optimization and the relative strength of the two spectra

pellicle in the gate beam path. The resulting minimum power transmission for gate pulse powers ranging from 225 to 440 mW is shown in Fig. 4.3b. A fit through 100% at 0 mW gate pulse power with a relative error of 3% confirms the linear scaling within this measurement uncertainty.

After upconversion, the following effects reduce the overall detection efficiency: the transmission properties of the SPF, the spatial overlap of the gate and sum-frequency beam and additional reflection and transmission losses of the remaining optics. First, we characterized the effect of the spectral filter that is used for dynamic range optimization. At the delay position of maximum EOS signal, we measured the sum-frequency spectrum behind the EOS crystal after a polarizer, rotated to transmit horizontally polarized light to suppress the strong gate pulse spectrum. A measurement with blocked MIR beam serves as the background. Figure 4.4 shows a measurement of the gate pulse spectrum (with orthogonal polarizer orientation) and of the sum-frequency spectrum with and without the SPF.

The full sum-frequency spectrum has a similar shape as the gate pulse, but is shifted by 40 THz (7.5 μm) which corresponds to the peak of the spectral response (see Fig. 4.9a). Of the generated sum-frequency photons at maximum signal, 60% are transmitted through the SPF at 912 nm.

Furthermore, photons are lost due to the imperfect spatial overlap of sum-frequency and gate beam. We calculate an overlap of 95% in the collimated beam, taking the sum-frequency focus size as the product of the almost perfectly matching MIR and NIR foci. Additional reflection losses at transport and polarization optics reduce the transmitted power to $\approx 87\%$. Taking into account Fresnel losses of the MIR and the sum-frequency beam at the detection crystal, $\approx 0.76\%$ of all MIR photons impinging on the EOS crystal arrive at the balanced photodiodes.

Minimum Detectable Photon Number

The minimum detectable photon number is limited by the detection noise floor, given by the balancing of the local oscillator. This noise floor is minimized with a combination of a $\lambda/2$ and a $\lambda/4$ plate. Figure 4.5 shows a typical balanced radio-frequency noise spectrum, with 140 μW of optical power of the local oscillator reaching each diode.

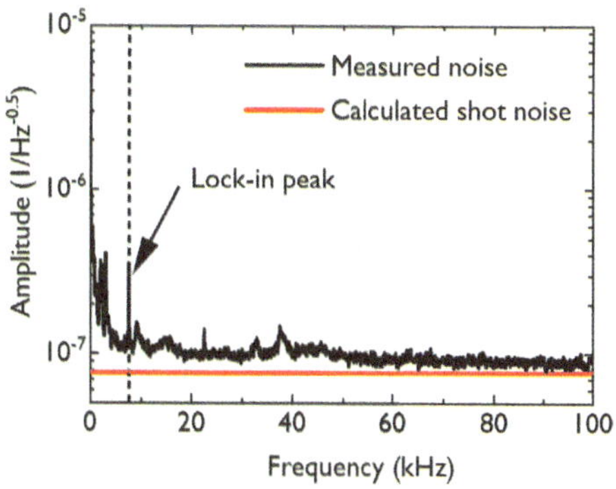

Fig. 4.5 Radio-frequency noise spectrum of the balanced local oscillator (black). Red: calculated shot noise. The dashed line indicates the lock-in detection frequency of 7.5 kHz. Adapted from ED Fig. 4.4a in [2]

The theoretical value for the shot noise limit was calculated with Eq. (2.47), for $\lambda = 900$ nm and $P = 0.140$ mW. The peak at the chopper-modulation frequency of 7.5 kHz is most likely caused by residual signal, because the noise-floor was measured at a delay away from optimum signal but without blocking the MIR beam. At the onset of the peak, the measured amplitude noise is a factor of 1.6 above the calculated shot noise. Possible reasons for this discrepancy are the gate pulse relative intensity noise (RIN) and the imperfect retardation of the waveplates. Consequently, a minimum average of $1.6^2 \approx 3$ sum-frequency photons are required in addition to the local oscillator photons arriving at the balanced detectors in the measurement time window at each delay point to surpass the detection noise floor (see Sect. 2.5).

4.2 Dynamic Range and Background-Reduced Measurements

For a given EOS signal strength and the balanced noise floor, measured with a blocked MIR beam, the time-domain amplitude dynamic range is calculated as the ratio of the two. This is shown in Fig. 4.6.

Signal and noise were filtered with a super-Gaussian filter in the frequency domain, limiting the spectral region to where the EOS signal is non-zero. For a 500-μm-thick GaSe crystal, the peak of the amplitude dynamic range is 5.4×10^6 for 2.4 ms (lock-in time constant of 183 μs with a 6-th order filter) measurement time per temporal element, (light red curve). The time-domain signal of the main pulse extends up to 2 ps, caused by the strongly limited detection bandwidth (see Fig. 4.8 and discussion below). After the main pulse, the signal stays two orders of magnitude above the noise floor, as the free induction decay from water molecules in the sample cell is measured. For the characterization of biological measurements, water often serves as a reference measurement. Therefore, the trace measured for water is usually subtracted from the actual measurement. The background for those measurements is thus calculated by subtracting two reference measurements, as shown by the dark red line.

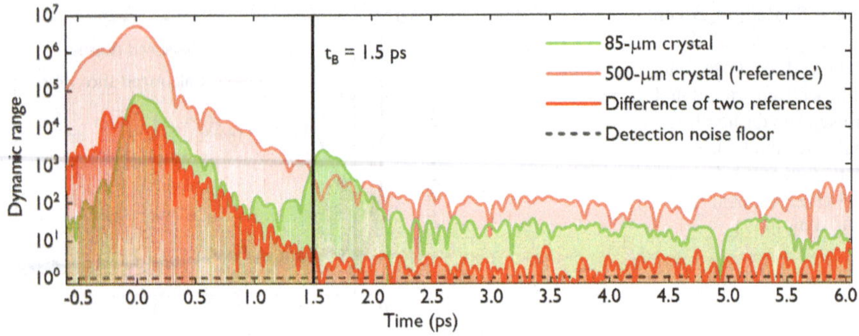

Fig. 4.6 Time-domain amplitude DR of EOS traces for 500 and 85-μm-thick GaSe crystals, also showing the difference of two subsequent measurements with the 500-μm-thick one. t_B: Time where the difference of two subsequent measurements reaches the detection noise floor. Adapted from Fig. 4.2a in [2]

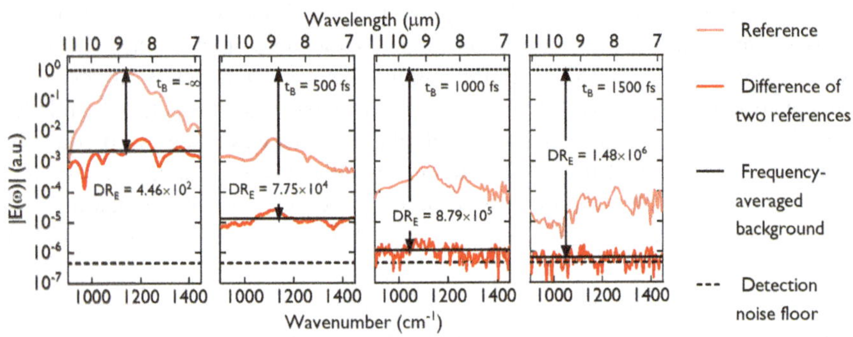

Fig. 4.7 Frequency-domain amplitude calculated from the time-domain traces for different cut-on times t_B. Light red: Fourier transform of a single measurement. Dark red: Fourier transform of the difference of two references. The effective dynamic range is limited by the remaining background for the difference measurement. Based on Fig. 4.2b in [2]

At $t_B = 1.5$ ps, this difference reaches the detection noise floor, thus opening a time window where the sensitivity of the free-induction-decay measurement is limited by the detector noise. The remaining background for referencing measurements with varying high-pass-time-filtering times t_B is illustrated in the frequency domain in Fig. 4.7.

From left to right, the cut-on time t_B for the Fourier transform is shifted from the beginning of the time-domain trace to 1.5 ps after the main pulse, as indicated in Fig. 4.6. For each case, the light red line shows the Fourier transform of a single measurement and the dark red line is the Fourier transform of the difference of two measurements. The effective noise floor for a spectroscopic measurement is given by the difference of the two references. Only in the last case, for $t_B = 1.5$ ps the noise floor is reached up to a factor of two, that is the limit when subtracting two noise traces. In this case, the full dynamic range can be exploited for measurements, not being limited by residuals from the reference measurement. This condition we refer to as a 'background-free measurement'. In order for this background-free time-window to start at earlier times t_B, or at the same time for a spectrally more complex solvent, better subtraction of two traces and/or a faster roll-off of the EOS trace is required. The green line in Fig. 4.6 shows the time-domain amplitude dynamic range for a 85-μm-thick detection crystal, peaking at 10^5. For this detection crystal, the pulse decays roughly twice as fast as for the 500-μm-thick one, as the detection bandwidth is more braodband (see spectral comparison in Fig. 4.8). However, a second peak occurs at 1.6 ps delay, corresponding to a reflection of the MIR pulse off the back surface of the GaSe crystal. This reflection will be discussed in more detail in Sect. 4.4.

In addition, the time-domain trace can be temporally compressed using tailored chirped mirrors for the MIR field [8]. This was applied here for the 85-μm-thick crystal, were a broad MIR frequency range is detected and the uncompensated dispersion is the dominating effect for the trace elongation. In addition, reducing the measurement time for a single trace can reduce the influence of drifts and therefore improve the subtraction of two traces. Methods to achieve fast delay-scanning of the gate pulse are asynchronous optical sampling (ASOPS) [12] or the use of an oscillating mirror [13]. An electro-optic detection system employing the first technique will be discussed in Sect. 4.6.

4.3 Bandwidth-Efficiency Trade-Off

Section 2.3 and Chap. 3 discussed the increase of the MIR to sum-frequency power conversion with the square of the length of the detection crystal. However, this also affects the phasematching bandwidth of SFG, resulting in a trade-off between the detection efficiency and bandwidth. While the aim discussed in Sect. 4.1 was to maximize the number of detected photons, the detection bandwidth for the 500-μm-thick EOS crystal is limited to a spectral width of 1.2 μm around 9 μm at the −10 dB level, shown as the black line in Fig. 4.8b.

The limited spectral coverage is detrimental for the analysis of samples with a broad distribution of absorption lines, like biological samples. Therefore, the measurements in [2] were taken in a broadband phasematching configuration with a 85-μm-thick crystal, at the expense of maximum detection efficiency. Figure 4.8a shows measurements of the MIR power depletion for different EOS crystal thicknesses and phasematching settings. The average minimum depletion in panel (a) reduces from 2% for the 500-μm-thick crystal to 0.3% for the 85-μm-thick one. Furthermore, the thinnest crystal shows the strongest interference fringes, as phasematching for DFG and temporal walk-off are less severe. The red and the blue lines show two different phasematching configurations for a 300-μm-thick crystal. Because of the more broadband phasematching for longer wavelengths, the depletion is larger in that case. The detection dynamic range and spectral coverage is illustrated in Fig. 4.8b for a MIR-average power of ≈ 1 mW at the detection crystal and a measurement time

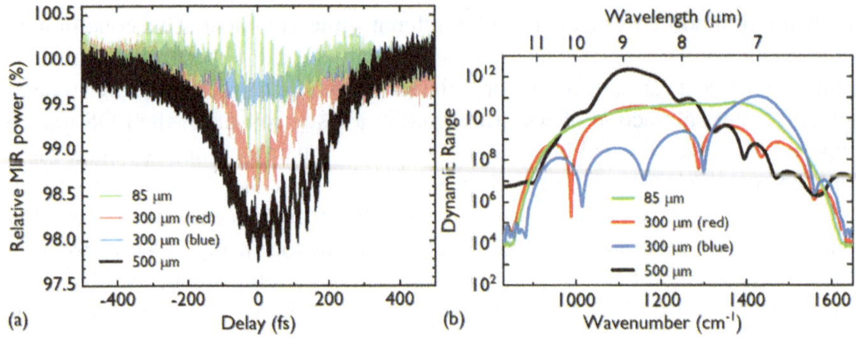

Fig. 4.8 Comparison of MIR power depletion and spectral intensity dynamic range for different EOS crystal thicknesses. **a** Relative MIR power transmission through the EOS crystal. **b** Frequency-domain intensity dynamic ranges

of 16 s for each thickness. From the 500-µm-thick to the 85-µm-thick crystal, the maximum intensity dynamic range decreases by almost two orders of magnitude, but the FWHM spectral coverage extends from 6.6 µm to 10.6 µm at the −20-dB-level. For the intermediate crystal thickness of 300 µm, two phasematching settings were tested. Optimum phasematching for the red curve is similar to the 500-µm-thick crystal, while the spectral response was tuned to peak at 7 µm for the blue line. This tunability allows for the selection of a spectral range of interest with locally increased detection efficiency.

Electric Field Reconstruction

EOS traces measured with the spectrally flat and broadband spectral response of the 85-µm-thick detection crystal closely resemble the MIR electric field. Remaining limitations to the spectral response, caused by phasematching, the gate-pulse duration or spectral filtering, can be corrected with a calculated IRF for the measured gate pulse (Fig. 4.2) and the given crystal geometry—$\theta = 55°$ and $\varphi = 0°$—as explained in Sect. 3.3. Figure 4.9 shows the simulated spectral and temporal IRF.

The spectral intensity IRF shown in panel (a) has a two-peak structure, with the FWHM width spanning from 7.2 to 8.6 µm and the long wavelength tail of the response extending to 10.6 µm at −10 dB. In the time domain, this corresponds to a gating window with an intensity-FWHM duration of 52 fs. With these results, the measurement was corrected for the spectral amplitude and phase response, resulting in the retrieved spectrum and field shown in Fig. 4.10.

The retrieved spectrum in panel (a) shows an increase in the spectral components in the spectral wings, that were suppressed by the IRF. Suppression of those wavelengths and propagation through the detection crystal also temporally compresses the EOS signal as compared to the retrieved field, see panel (b). Overall, both waveforms are similar in length and shape.

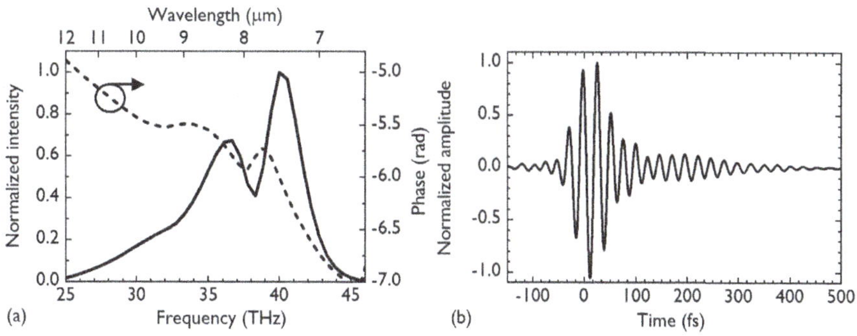

Fig. 4.9 Calculated EOS IRF for 1-μm gate pulses in a 85-μm-thick GaSe crystal rotated to $\theta = 55°$ and $\varphi = 0°$. **a** Intensity and phase of the frequency-domain IRF. **b** Normalized amplitude of the time-domain IRF

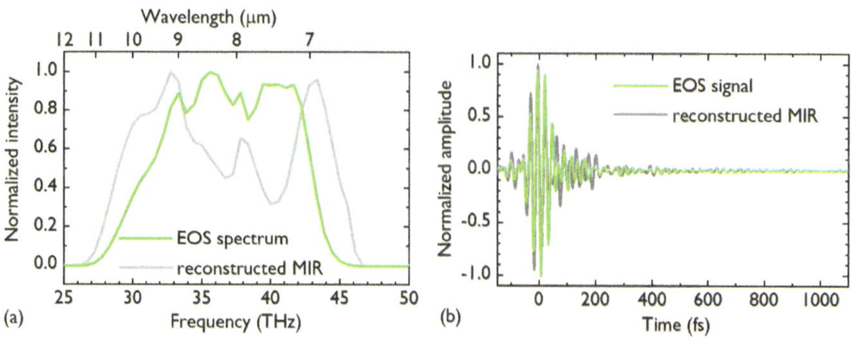

Fig. 4.10 Comparison of EOS measurement and reconstructed MIR field for a 85-μm-thick EOS crystal. **a** Measured EOS and reconstructed MIR spectrum after correction with the calculated IRF. **b** Measured EOS trace and retrieved field, corrected for the calculated spectral response

While knowledge of the exact field is not necessary for the linear spectroscopic measurements performed with this EOS detection setup, a more realistic estimation of the field allows us to investigate the influence of changes to the detection geometry, e.g., the crystal thickness or spectral filters, on the measured traces. In addition, the effects of chirped mirrors [8] on the temporal compression of the measured trace or the MIR field can be understood.

4.4 Multiple Reflections from GaSe

As discussed in Sect. 2.4, GaSe stands out with its unique phasematching properties, relatively large damage threshold and high nonlinear coefficient. Therefore, it is our crystal of choice for EOS detection. However, it cannot be cut at a certain phase-matching angle, as the two-dimensional crystal layers are only weakly bound and the material easily breaks along cleavage lines. This property also makes anti-reflection coatings challenging. Both MIR generation and detection in GaSe therefore suffer from power losses and the visibility of internal reflections of the gate and MIR beams. For EOS, reflections of the MIR beam off the back surface of the detection crystal appear at delays behind the main pulse, in the same time window were the free-induction decay of molecules is analyzed in spectroscopic measurements, thus limiting the background-free measurement time window. For the 85-μm-thick crystal, this is shown in Fig. 4.6.

Up to now, anti-reflection coating of GaSe was demonstrated in [14] for 800-nm wavelength with magnetron sputtering of SiO_2 and SiO_2/Si_3N_4 layers, leading to a higher THz generation efficiency, as less driving pulse power is reflected off the front surface. Furthermore, Bushunov et al. and Hagner et al. [15, 16] show that the fabrication of anti-reflection micro-structures can be used to increase the MIR transmission of GaSe crystals. However, the effect of the microstructures on the crystal damage threshold still has to be investigated. Because the structures are on the scale of the gate pulse wavelength, they act as a grating, which makes the crystals promising for IPDFG, but so far not for EOS, where the gate beam (polarization) properties are crucial.

The use of thick detection crystals can shift the appearance of the first echo to beyond the time window of interest. Furthermore, this can reduce the detection of the reflection because of spatial walk-off. However, the thick crystals limit the detection bandwidth, as discussed above for 1-μm gate pulses. The following sections show EOS using longer gate pulse wavelengths with a broader phasematching bandwidth and thus thicker crystals. Here, three different approaches to suppress the multiple reflections which were tested within the scope of this thesis, are discussed.

1. In a collaboration with M. Knorr and Prof. R. Huber at the University of Regensburg, we tried the approach they present in [17]. GaSe is optically contacted to a poly-crystalline diamond substrate *via* van-der-Waals bonding upon evaporation of a droplet of methanol. At 8-μm wavelength, the refractive index of diamond matches the extra-ordinary refractive index at normal incidence of GaSe. One difference to their approach is that we use crystals with $\geq$ 85 μm thickness to maximize the signal strength, while they have almost single-layer, $<$ 10-μm-thin samples. We observed that only one or a few of the GaSe layers contacted to the diamond substrate, while the rest peeled off. We therefore first contacted 5-μm-thin GaSe to diamond and then attached a second, 60-μm-thick crystal to the other side of the thin GaSe. The resulting EOS measurements are shown in Fig. 4.11. Panel (a) shows the time-domain trace, with a reference measurement for a 85-μm-thick GaSe crystal for comparison. The frequency-domain spectra and the noise floor are shown in panel (b). From the time-domain traces, no reduction of the multiple reflection is observable, indicating that no optical contact was formed. The modulations in the frequency-domain plots for the 'contacted' crystal are most likely caused by the slightly different phasematching angle. Furthermore, the noise floor for the contacted crystal is three orders of magnitude larger than for the reference measurement. We observed a strong dependence of the balanced

noise floor on the gate-beam position on the nonlinear material and therefore assume that the poly-crystalline diamond substrate caused depolarization of the gate pulses. For future experiments we therefore suggest to use single-crystalline diamond to mitigate the depolarization effect.

2. Similar to the application in [18], Arsenic Sulfide (As_2S_3) was investigated as a material with matching refractive index to GaSe. We attempted to contact a 500-μm-thick diamond plate to a GaSe sample by using a layer of Arsenic Sulfide in between. The material was applied in liquid form, solved in propylamine. However, when the solvent evaporated to form the contact, a crystalline Arsenic Sulfide structure formed, which strongly distorted the NIR beam profile. The optimization of the drying process has to be investigated in more detail for future attempts.
3. Optical glue with a refractive index matching that of GaSe is an additional option to reduce the strength of the reflection or shift it to later delays. The glue with the highest available refractive index is NOA 170 (from Norland), with $n_{\mathrm{NOA170}} = 1.7$. We attempted to reduce the MIR reflection by gluing a 500-μm-thick, single-crystalline diamond plate to the back-surface of the GaSe crystal. A comparison of a reference measurement and an EOS trace for the glued crystal is shown in Fig. 4.12.
   Due to the limited aperture of the glued crystal, the phasematching angle was reduced to $\theta = 46°$ instead of $\theta = 52°$ for the reference measurement. In addition, despite the same nominal thickness, the crystal used for gluing was more than twice as thick, as can be seen from the MIR reflection shifted to longer delays. Taking into account all Fresnel reflections and no spatial effects, a reduction of the MIR reflection in the field of 70% was expected. The measured value shows the opposite behaviour, even though a reduction of the spatial overlap of the inner reflection with the gate pulse is expected for the thicker crystal. We therefore assume that no optical contact between the substances was formed. Investigating the inner reflection of the s-polarized gate-beam, which appears before the main signal, should increase the visibility of potential improvements, because of the increased change in reflection. Suggesting an alternative way to use the optical glue, we propose to attempt to achieve spatial separation of the multiple reflections by forming a lens with a droplet of glue. This approach is still to be investigated in more detail.

Furthermore, LGS, which is widely used for IPDFG with 1-μm driving pulses, was tested as an alternative detection crystal. This material can be anti-reflection coated and has a broader phasematching bandwidth and higher damage threshold than GaSe (see temporal walk-off in Table 2.1), which allows for the use of approximately five times thicker crystals while maintaining the detection bandwidth. Figure 4.13 shows the resulting frequency-domain intensity dynamic range for 500-μm-thick LGS, compared to that for 85-μm-thick GaSe.

The signal strength for both crystals was comparable when using 390 mW of gate pulse power for GaSe and 620 mW for LGS, with the increased power approximately compensating for the lower nonlinear coefficient. Note that the higher damage thresh-

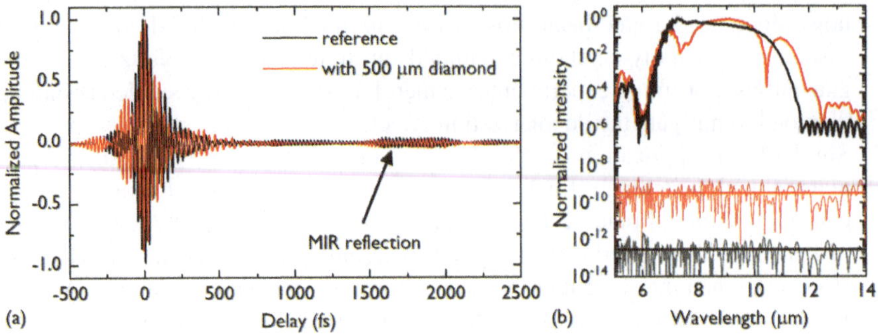

Fig. 4.11 EOS measurements with GaSe contacted to diamond. **a** Comparison of the EOS traces. **b** Comparison of the intensity spectra and noise measurements

Fig. 4.12 Comparison of EOS traces measured with a bare GaSe crystal (black) and with a 500-μm-thick diamond plate glued to the back surface of the crystal (red). The numbers indicate the strengths of the visible MIR reflections

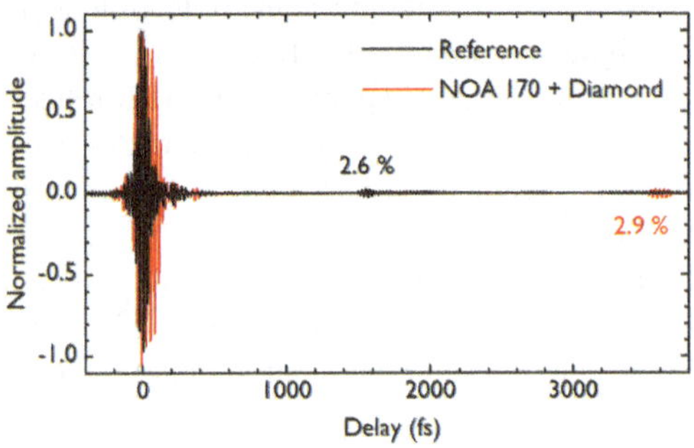

Fig. 4.13 Frequency-domain intensity DR for EOS with a 500-μm-thick LGS crystal. Comparison to the dynamic range measured with 85-μm-thick GaSe

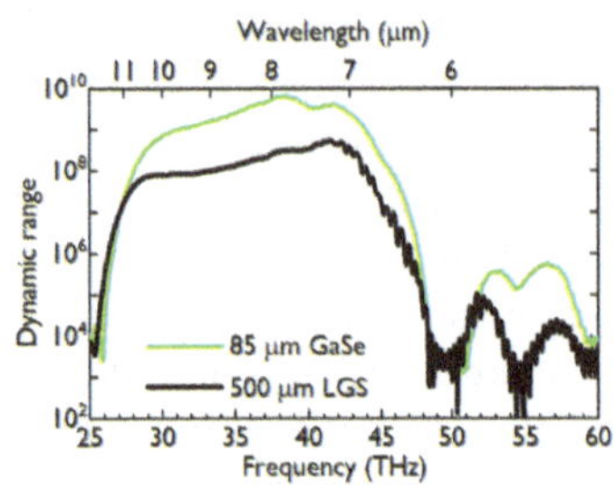

old of LGS would allow the use of even higher powers and thus increase the signal strength. However, the balanced noise floor was a factor of 2.7 worse in amplitude. This was caused by the bi-axial nature of the crystal, which lead to severe depolarization of the local oscillator. Therefore, the intensity dynamic range is a factor of 23 lower at 8 μm wavelength for the LGS crystal. To reverse the depolarization, we suggest to use a second LGS crystal in opposite orientation. This approach still needs to be tested experimentally.

In conclusion, attempts to suppress the multiple reflections of the MIR beam at the GaSe surfaces by optical contacting of several materials were tested, however so far with limited success. Using LGS as a detection crystal is promising in terms of signal strength and spectral coverage. However, uncompensated depolarization of the local oscillator due to birefringence limits the balanced noise floor. Currently, for the analysis of spectroscopic measurements performed in our working group, the multiple reflections are suppressed numerically.

4.5 Application of High-DR EOS: Waveform Stability Analysis

In Sect. 4.2, electro-optic detection with a time-domain amplitude dynamic range of more than six orders of magnitude was demonstrated. Besides the application for spectroscopic measurements, as described in [2], this high sensitivity allows for the precise characterization of the temporal stability of the EOS traces and thus of the MIR waveforms with respect to the gate-pulse reference. On the one hand, this provides information on the remaining timing jitter after passive optical-phase stabilization *via* IPDFG. On the other hand, the quantification of the remaining jitter and its time scales provides insight on how well and over which time scales consecutive measurements can be averaged to increase the overall measurement sensitivity.

For an analysis of the stability of the investigated waveforms, we measured the amplitude extrema and zero-crossing position jitter of the EOS traces. To draw conclusions about the jitter of the actual electric fields, a theoretical model was developed, that propagates the laser-fronted RIN through all nonlinear processes involved in the generation and detection of the waveforms. The experimental results of this study are discussed in detail in [3] and were presented in [19]. Here, the focus is on the measurement sensitivity and the correspondence between the fluctuations measured with EOS and those of the electric field.

The waveform stability was measured by setting the nominal gate pulse delay to the positions of various zero-crossings and extrema of the EOS trace at a time, recording the signal for millions of pulses and analyzing the relative standard deviation for the extrema and the absolute variations at the zero-crossing positions.[3] The detection crystal was 85-μm-thick GaSe, thus corresponding to the bandwidth-maximized measurement configuration shown in Fig. 4.10b, where the measured trace closely resembles the input field. At the extrema, the signal variation is directly proportional to the amplitude fluctuations of the EOS trace. For the zero-crossings, the signal variation $\Delta\hat{S}$ was transferred to a jitter in the position of the zero-crossing *via* the slope of the optical waveform: With the instantaneous angular frequency ω and signal amplitude $\hat{A}$, the oscillating signal $\hat{A} \times \sin(\omega\tau)$ can be approximated with the 0th-order term of the Taylor expansion: $\Delta\hat{S}(\tau) \approx \hat{A}(\tau) \times \omega(\tau) \times \Delta\tau$.

Figure 4.14 summarizes the results of those amplitude and zero-crossing position jitter measurements for a chirped and a compressed MIR wave.

[3] The waveform stability measurements and data analysis were performed by S. A. Hussain and W. Schweinberger. T. Buberl helped with the data analysis.

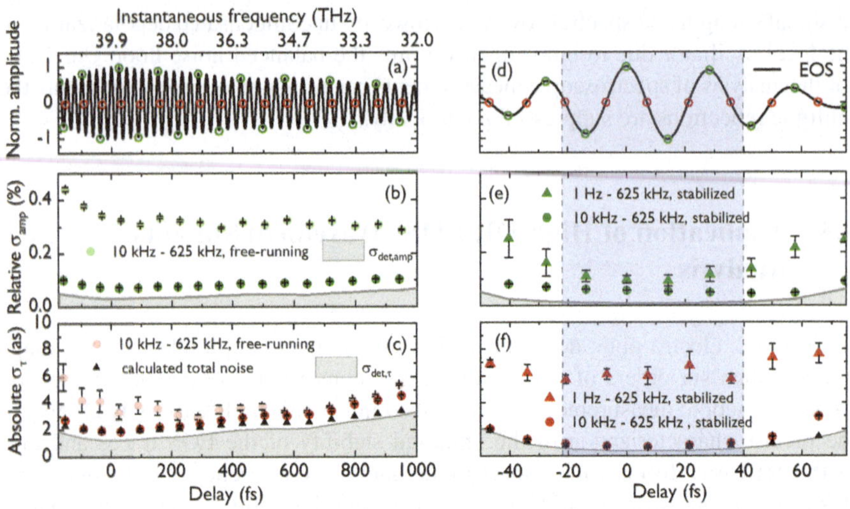

Fig. 4.14 Waveform stability measurements for a temporally chirped (**a–c**) and compressed MIR waveform (**d–f**). **a** EOS trace of the chirped MIR waveform, red and green circles indicating delay points for zero-crossing and amplitude jitter measurements, respectively. **b** Measured amplitude jitter, with (dark green) and without (light green) power stabilization of the driving pulses of IPDFG. **c** Measured timing jitter, with (dark red) and without (light red) power stabilization. **d** Temporally compressed EOS trace, red and green circles indicating delay points for zero-crossing and amplitude jitter measurements, respectively. **e** Measured amplitude jitter for the radio-frequency ranges 1 Hz–625 kHz (triangles) and 10–625 kHz (circles). **f** Measured timing jitter for the two radio-frequency ranges. The grey shaded areas show the calculated measurement sensitivity. Figure adapted with permission from the author from [3]

Panel (a) shows a measured EOS trace of the chirped waveform, obtained from the compressed waveform in panel (d), by adding a 5-mm-thick CaF_2 substrate to the MIR beampath. From left to right, the instantaneous frequency decreases monotonically, corresponding to a wavelength change from 7.2 to 9.3 μm. The red and green circles indicate the delay positions at which the waveform stability was measured. For the chirped pulse, the fluctuations were analyzed in the radio-frequency range from 10 to 625 kHz, with and without power stabilization of the IPDFG driving pulses with the acousto-optic modulator (AOM) (see the setup description in Fig. 4.1). We chose this radio-frequency range for the analysis, because the interferometer length between MIR field and gate pulse is assumed to be constant, avoiding the measurement of drifts in the temporal delay between the pulses. The upper frequency-limit is determined by the measurement electronics.

Sensitivity of the Waveform Stability Measurements

The plots of the measured fluctuation values also show the measurement sensitivity for the amplitude and zero-crossing fluctuations as the grey areas. This sensitivity was determined from the time-domain amplitude dynamic range: At the maximum of the compressed pulse, we measured the average amplitude value of the signal,

$\bar{S}_{\max}(0) = 4.8\,\text{V}$. With blocked MIR, the standard deviation of the noise floor was $3 \times 10^{-4}\,\text{V}$, resulting in an amplitude dynamic range of 1.6×10^4. Therefore, the minimum measurable amplitude change, calculated as the inverse of the dynamic range, is $\sigma_{\text{det,amp}}(0) = 6 \times 10^{-5}$ at the delay position of maximum of the EOS trace envelope ($\tau = 0$). When varying the delay τ to positions away from this maximum, the minimum detectable change increases with the inverse signal amplitude $\hat{A}(\tau)$: $\sigma_{\text{det,amp}}(\tau) = \sigma_{\text{det,amp}}(0)/\hat{A}(\tau)$. As the timing jitter is also measured *via* the signal change, its detectable minimum is related to the smallest measurable amplitude change with the instantaneous frequency ω of the compressed pulse: $\sigma_{\text{det},\tau}(0) = \frac{\sigma_{\text{det,amp}}(0)}{\omega} \approx 0.5\,\text{as}$. The influence of the variation of the slope of the EOS signal, caused by the amplitude jitter, that is used for the transformation of the amplitude to the timing jitter, is on the order of 0.01 as and therefore considered negligible. As the signal strength is reduced for the chirped pulse, the measurement sensitivity is decreased in that case.

Measurements Results

In the free-running case (light green), the amplitude fluctuations increase from 0.3% at 9.3 μm to 0.4% at 7.2 μm, as shown in Fig. 4.14b. The data points are the average values for five consecutive measurements and the error bars correspond to the standard deviation. With the stabilization switched on (dark green), the amplitude fluctuations are approximately constant at 0.1% at all wavelengths. Without power stabilization, the positions of the zero-crossings jitter with an amplitude between 4 as and 5 as over the full optical wavelength range (light red dots in panel (c)). Switching on the AOM leads to a decrease of the fluctuations at the shortest wavelengths to 3 as (dark red). For both the amplitude and zero-crossing fluctuations, the power stabilization has the strongest effect for the shortest wavelengths. This is due to the error signal of the stabilization being generated in the short-wavelength/high optical frequency wing of the NIR driving spectrum, that contributes the most to the generation of high MIR frequencies/short wavelengths.

For the compressed pulse (panel (d)), both the amplitude and timing jitters values are calculated from the measured signal changes for two different radio-frequency ranges with the power stabilization switched on. This allowed us to investigate the contributions of low-frequency noise to the waveform stability. In the range from 1 Hz to 3 kHz, the data was corrected with position data recorded for the gate pulse delay with interferometric delay tracking (IDT), compensating for drifts in the interferometer length (details in Hussain [3] and Schweinberger et al. [9]). Within the FWHM of the compressed EOS trace, the average amplitude fluctuations are 0.1% in the frequency range from 1 Hz to 625 kHz, see Fig. 4.14e. Selecting only the higher frequencies from 10 to 625 kHz reduces the average amplitude fluctuations within the FWHM of the trace to 0.1%. In this range, drifts of the interferometer are negligible, thus making the position correction obsolete. The average zero-crossing jitter in the full and limited frequency range is 6 as and 0.8 as, respectively, see Fig. 4.14f.

Comparing the measured timing jitter values and the measurement sensitivity shows that the two are close together and that the shape of the measured values is similar to that of the noise floor. In order to show whether the measurement was fully dominated by the noise floor or additional laser noise is detectable, we assumed the total jitter to be given by the sum of the uncorrelated detector and laser noise. To demonstrate the effect of purely increased detector noise, we assume a constant laser noise, i.e., at the lowest measured value at maximum signal $\sigma_{\text{laser},0}$. This results in: $\sigma_{\text{tot}}(\tau) = \sqrt{\sigma_{\text{det}}(\tau)^2 + \sigma_{\text{laser},0}^2}$. This data is shown as the black triangles in Fig. 4.14c, f. The deviations from the calculated and measured fluctuations shows that the increase of the measured fluctuations away from the maxima of the traces is not solely caused by the decreased sensitivity, but that the traces fluctuate more at those delays.

In conclusion, the stability of waveforms generated *via* IPDFG was characterized by measuring the variations of the EOS signal. This resulted in amplitude and timing jitters on the order of 0.1% and 1 as, respectively. Optical-wavelength dependent effects were illustrated by measuring with and without the power stabilization of the driving pulses for IPDFG.

Theoretical Model and Simulations

The waveform stability measurements were performed with a 85-μm-thick GaSe crystal. Although Fig. 4.10 shows the similarity of the measured trace and the input field, attosecond-level fluctuations could be averaged due to propagation effects in the detection crystal, i.e., the temporal walk-off of detected field and gate pulse. Furthermore, amplitude fluctuations of the gate pulse directly affect the amplitude fluctuations of the measured trace, as no normalization to the gate pulse intensity is performed when recording the electro-optic signal.

For understanding the connection between the measured fluctuations and the actual jitter of the MIR fields, we developed a theoretical model, simulating the chain of nonlinear effects in the experimental setup.[4] This model is sketched in Fig. 4.15.

The simulation input (see Fig. 4.15a) is the experimentally determined laser oscillator output, a sech^2-shaped soliton pulse with 220 fs FWHM duration, characterized with an autocorrelation measurement. Figure 4.15b depicts the measured RIN spectra with and without the power stabilization of the NIR pulses with an acousto-optic modulator. Three stages of Herriott-cell-type spectral broadening in fused silica (Fig. 4.15b) are simulated with the nonlinear phase shift φ_{Kerr} for self-phase modulation [20] in the time domain:

$$\begin{aligned} \varphi_{\text{Kerr}} &= -4n_2 I k_0 d \\ A_{\text{after}} &= A_{\text{in}} e^{i\varphi_{\text{Kerr}}} \end{aligned} \tag{4.1}$$

with the complex amplitudes A_{in} and A_{after} for before and after the nonlinear interaction, respectively. For fused silica, the nonlinear coefficient is $n_2 = 3.5 \times 10^{-20}\,\text{m}^2/\text{W}$ [20] and we use substrates with a thickness of $d = 6.35\,\text{mm}$. The

[4] M. Högner helped developing the theoretical model, performed parts of the simulations and crucially contributed to understanding the results.

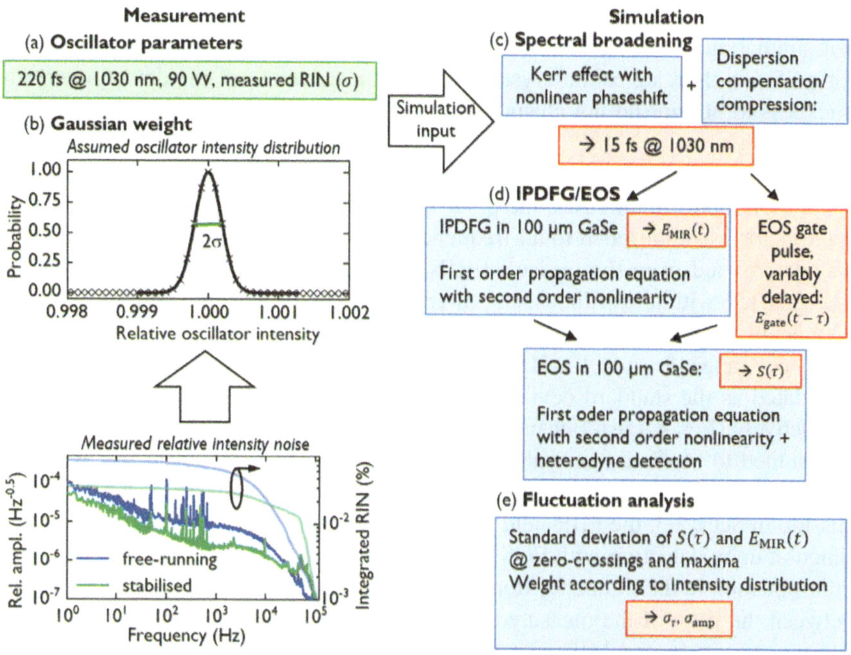

Fig. 4.15 Sketch of the theoretical model for the waveform stability analysis. **a** Simulation input: Oscillator parameters and measured relative intensity noise as shown in **b**; stabilized: with power stabilization using an acousto-optic modulator (AOM). **c** Simulation of the spectral broadening of the oscillator output, resulting in 15-fs NIR pulses. **d** Simulation of IPDFG and EOS, both in GaSe, with the spectral coverage matching the experimental parameters. Simulations are performed for several oscillator output powers, assuming a Gaussian distribution of the intensity values (see panel **b**) with the width corresponding to the measured RIN value from **a** (σ). **e** Fluctuations are calculated as the standard deviations of the simulated amplitude and zero-crossing values for MIR fields and EOS traces, weighted according to **b**

wavenumber at the central frequency ω_0 is given as $k_0 = \omega_0/c$. From the complex time-domain amplitude A_t, the intensity is calculated as $I = \frac{n|A_t|^2}{2c\mu_0}$. The linear propagation over the broadening cell chirped mirrors considers their phase φ_{HC} (calculated as in (2.4)), and reflectivity, as provided by the mirror design.

After the spectral broadening, the temporal compression with chirped mirrors is implemented analogously to the chirped Herriott-cell mirrors with a linear phase shift. We adjusted the peak intensity in the broadening medium to achieve Fourier transform limited pulse durations after each of the three stages that match the experimental values, i.e., 80, 43 and 15 fs, respectively. For the simulation of IPDFG (Fig. 4.15d), we implemented the first-order propagation equation with the second-order nonlinearity, as described in Sect. 2.3, for a 100-µm-thick GaSe crystal as a simple model system. This generates a spectrum spanning 27–47 THz at −30 dB, which agrees well with the experimental IPDFG results. This field is optionally chirped in time to extend over more than 1 ps, by adding the spectral phase of 5-mm-thick CaF_2. For simulating EOS detection, the theoretical model described in Sect. 2.3 is used for a 85-μm-thick detection crystal. Note that a reduction of the detection crystal thickness to 30 μm did not change the qualitative results of the simulation results presented here.

The full chain of nonlinear processes was simulated for various intensities of the soliton input. In all cases, the pulse shape was kept identical, thus only a constant factor was multiplied to all frequencies. The variation of the oscillator power was determined from the measured RIN as shown in the radio-frequency spectra in Fig. 4.15b with and without the power stabilization. All simulation results were then analyzed according to the measured integrated RIN value of $\approx 0.02\%$ for the frequency range down to 10 kHz. The timing-jitter and amplitude fluctuations were calculated as the standard deviation of the zero-crossing times and field extrema, respectively (Fig. 4.15e). Because the intensity distribution of the oscillator output is assumed to be Gaussian with a width corresponding to the measured RIN, the amplitude and zero-crossing values were weighted accordingly. Referring to the discussion in Sect. 2.3, the MIR field is convoluted with a gate or instrument response function upon detection with EOS. This time-domain convolution corresponds to a multiplication in the frequency domain and suggests to analyze the correspondence between the jitter of the measured trace and the input fields also in the frequency domain. For a chirped MIR waveform, with its individual frequencies mapped to different times, this frequency-domain behavior can be analyzed with a time-domain measurement. Simulation results for the time- and frequency-domain jitter of the chirped waveform are depicted in Fig. 4.16.

Panel (a) shows the normalized modulus squared of the frequency-domain complex amplitude for the MIR, IRF and EOS. Following the description of the IRF in Sect. 3.3, the complex spectral amplitude of the EOS signal results from a multiplication of the MIR complex amplitude with that of the IRF in the frequency domain. As a result, and using the amplitude and phase representation for the frequency-domain fields introduced in Eq. (2.1), the following relations hold for the relative fluctuations of the EOS spectral amplitude and the absolute phase jitter:

$$\Delta A_{\mathrm{EOS}} = \Delta A_{\mathrm{MIR}} + \Delta A_{\mathrm{IRF}} + \Delta A_{\mathrm{MIR}} \Delta A_{\mathrm{IRF}} \tag{4.2}$$

$$\Delta \varphi_{\mathrm{EOS}} = \Delta \varphi_{\mathrm{MIR}} + \Delta \varphi_{\mathrm{IRF}} \tag{4.3}$$

The instrument response jitters due to fluctuations of the gate pulse. *Via* the dependence of the spectral broadening via self-phase modulation on the input pulse intensity, the temporal and spectral shape of the compressed gate pulses varies with the oscillator output power.

In panel (b), the simulated frequency-domain relative amplitude fluctuations are illustrated. For the MIR field, those are on the level of 0.05%. The amplitude fluctuations of the IRF vary between 0.1 and 0.25%, most likely caused by variations of the fluctuations of different components in the gate-pulse spectrum, which contribute most to the detection of the different MIR frequencies. Adding the fluctuations of

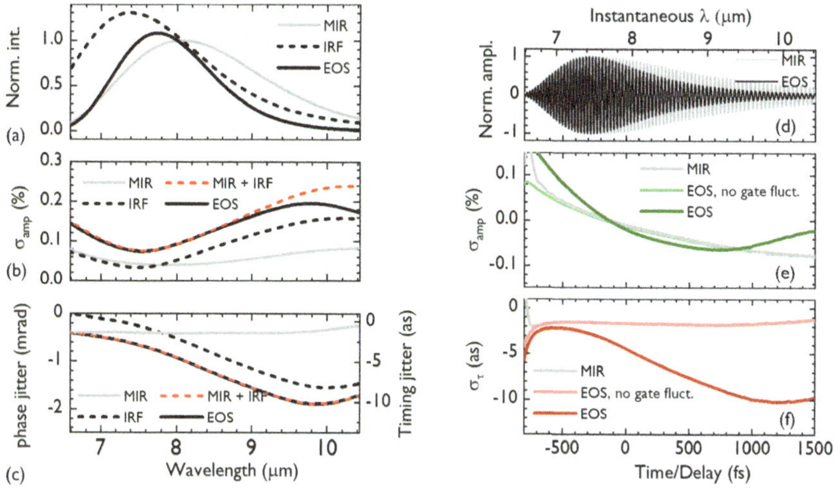

Fig. 4.16 Simulated jitter values for a chirped MIR waveform in the time and frequency domain. **a** Intensity spectra of MIR field (grey), EOS trace (black) and frequency-domain IRF (black, dashed). **b** Frequency-domain relative amplitude jitter for MIR (grey), EOS (black) and IRF (black dashed). For comparison, the red dashed line shows the sum of the MIR and IRF jitter. **c** Phase jitter for each case, also given as a timing jitter, calculated using the instantaneous wavelengths. **d** Simulated chirped MIR field (grey) and EOS trace (black). **e** Relative amplitude fluctuations for MIR field (grey) and EOS trace, with (dark green) and without (light green) gate pulse fluctuations. **f** Respective zero-crossing jitters

the field and the instrument response yields, in good approximation—except for low intensity values—the EOS amplitude fluctuations. Panel (c) shows the wavelength-dependent phase jitter. Because a phase shift of 2π corresponds to a time change of one oscillation period, the phase jitter $\Delta\varphi$ can be transferred into a timing jitter $\Delta\tau$, as shown on the right axis. The two are related with the instantaneous frequency ω *via* $\Delta\tau = \frac{\Delta\varphi}{\omega}$.

The MIR phase jitter is below 0.5 mrad over the full wavelength range, corresponding to timing jitters below 2.5 as. For the IRF, the timing jitter increases from 0.5 mrad at 7 μm to 2 mrad at 10 μm. Adding the jitters of the field and the IRF again yields the variations of the simulated signal. Furthermore, the IPDFG driving pulses and gate pulses are derived from the same source, suggesting that their jitters correlate. It shall be noted though that a dependence of the wavelength dependent jitter values on the configuration of the broadening stages was observed in the simulations: Driving the compression stages in the positive or negative dispersion regime leads to a different noise behavior [21]. Furthermore, wavelength dependent effects of the NIR power stabilization are not considered in the simulations. Discrepancies between the experimental and simulated broadening stages are therefore expected to lead to differences in the simulated fluctuation values.

For the chirped MIR waveform (panel (d)), the frequency-domain phase jitter (panel (c)) is directly comparable to the timing jitter (panel (f)), owing to the mapping

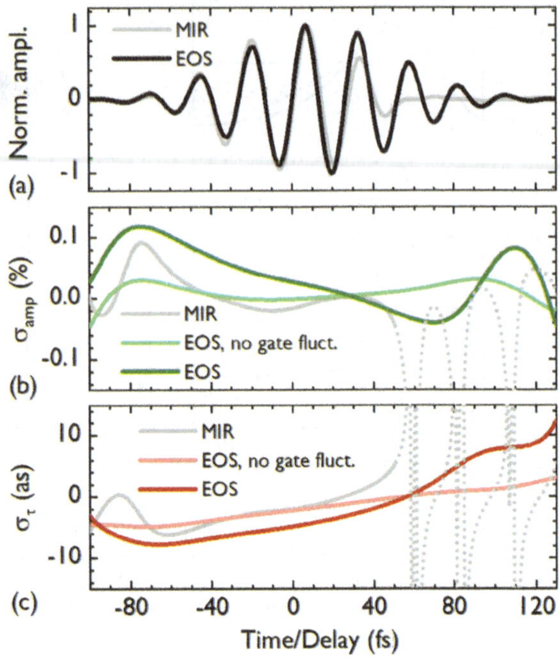

Fig. 4.17 Waveform stability simulation results for a compressed MIR field. **a** Compressed MIR field (grey) and EOS trace (black), the temporal offset is caused by walk-off in the nonlinear crystal. **b** Amplitude jitter of the compressed MIR field (grey) and EOS trace with (dark green) and without (light green) gate pulse fluctuations. **c** Corresponding timing jitter for the fields and EOS traces

of the spectral components to different temporal positions because of the wavelength-dependent group delay of CaF_2. The amplitude fluctuations in panel (e) do not match the behaviour in the frequency domain, most likely because of interference of the various spectral components and the influence of the spectral phase fluctuations on the time-domain amplitude jitter. Panels (e) and (f) also show the fluctuations of the simulated traces without gate pulse fluctuations, that means, with the simulated EOS detection using the same gate pulse for each of the varying MIR fields. In this case, the trace fluctuations in both amplitude and phase match the field fluctuations. Therefore, we conclude that the nonlinear detection process itself does not alter the time-domain fluctuations, despite temporal walk-off of the gate pulse and MIR field. However, fluctuations of the gate pulse cause a jitter of the instrument response, that in turn influences the shape of the measured fluctuations. The influence of the amplitude fluctuations of the gate pulse could be avoided by measuring the EOS signal normalized to the total local oscillator power reaching the balanced diodes.

Because the spectral response of the detection does not depend on its phase, compressing the investigated field in time does not change the behavior of the frequency-domain fluctuations. However, the differently fluctuating wavelengths are now interfering and therefore change the temporal jitter of the waveform. The simulated amplitude and timing jitter values for the compressed trace are shown in Fig. 4.17.

Panel (a) shows the compressed simulated trace and the MIR field. The two were temporally aligned at their maxima for better comparability. The relative amplitude fluctuations in panel (b) are below 0.1% for the field and the EOS trace, both with and without gate pulse fluctuations. At three delays between 40 and 120 fs (indicated by the dotted line), the amplitude jitter of the MIR field diverges (the same holds for the timing jitter). This effect ist caused by low-amplitude oscillations, which are not showing clear zero-crossings and therefore distort the jitter analysis. Because of the limited spectral coverage, the EOS trace extends longer in time, which avoids the occurrence of divergences for the considered delay window.

Panel (c) shows the timing jitter of the simulated fields and EOS traces, which have average values of 1.9 and 3.7 as within the FWHM of the field and trace, respectively. For both amplitude and timing jitter, their fluctuation levels are comparable, with the increase in the EOS timing jitter attributed to the fluctuations of the instrument response. Furthermore, the simulated jitter values are comparable to the ones determined experimentally.

In addition, the theoretical model allows for extrapolation of the jitter values to various RIN values of the oscillator front-end. The RIN for a shot-noise limited, 90-W output of the laser oscillator at 1030 nm wavelength and in the radio-frequency range from 10 to 625 kHz amounts to $\approx 5.1 \times 10^{-8}$. This value is approximately three orders of magnitude below the measured RIN in Fig. 4.15b and would correspondingly result in three orders of magnitude lower jitter values for amplitude and timing. In that case, the shot-noise limited RIN of the generated MIR radiation ($\approx$ 50 mW of average power at 8 μm wavelength) would however be dominating the amplitude fluctuations.

In conclusion, the fluctuations measured with EOS correspond to the sum of the input field and instrument response fluctuations in the frequency domain, where the complex spectral amplitude of the EOS signal is the result of a multiplication of the field and IRF complex spectral amplitudes. Because of the interference of different spectral components, this relation is not straightforward in the time domain for a compressed input field. Furthermore, it is suggested that the fluctuations of the NIR pulses that both drive the IPDFG and act as gate pulses in detection strongly depend on the configuration of the pulse compression scheme and that wavelength-dependent effects from the power stabilization affect the exact behavior of these fluctuations. The fluctuation levels between simulations and experiment agree qualitatively for comparable RIN. Furthermore, despite temporal walk-off between the gate pulse and the infrared field in the thick detection crystal, averaging of the field fluctuations at different delay values does not significantly alter the simulated fluctuation values of the EOS trace with respect to the input field. This makes high-dynamic-range EOS a powerful technique to investigate fluctuations of the electric field of light on a sub-cycle time-scale, where the measured fluctuation values correspond more closely to the electric-field fluctuations for lower gate-pulse fluctuations. The as-level jitter enables linear spectroscopy with high precision and allows for accurate tracking of the mutual delay between two pulse-trains for dual-oscillator scanning [22]. Using a detection geometry with a more broadband frequency response (i.e., with thinner detection crystals and/or longer gate-pulse wavelengths), in order for the EOS trace to resemble the MIR field more closely, this stability benefits the sensitive time-domain investigation of nonlinear effects [23–26]. For our experimental setup, a further reduction of the fluctuations would require a lower RIN of the laser frontend or a compression scheme with taylored noise characteristics for a reduction of the amplitude noise [27]. For a shot-noise limited laser oscillator output, the timing and amplitude jitter values are expected to be ≈ 3 orders of magnitude lower than the fluctuations measured here.

4.6 EOS with 1550-nm Gate Pulses

While the EOS results with the 1-μm gate pulses described in the previous section were unprecedented in terms of detection efficiency and already greatly benefit the sensitivity of spectroscopic measurements, both the MIR generation and detection were limited in efficiency and bandwidth due to crystal damage thresholds and phasematching limitations. Benefits from increasing the gate pulse wavelength were discussed with simulation results in Sect. 3.4. Erbium-fiber lasers with a central wavelength of 1550 nm are well-established as sources for gate pulses for broadband EOS in GaSe [28, 29]. However, they suffer from limited peak power, thus limiting the conversion efficiency for SFG of MIR and gate pulses. Here, phasematching benefits from using 1550-nm gate pulses in comparison to the 1030-nm-based EOS are demonstrated experimentally.

The infrared-field resolved detection system studied here, is based on the same Yb:YAG thin-disk technology as discussed in Sect. 4.1 and is sketched in Fig. 4.18.[5]

In short, an Yb:YAG thin-disk oscillator generates pulses at 1030 nm central wavelength with 100 W of average power and 200 fs FWHM pulse duration. Two Herriott cells and chirped mirror compressors shorten the pulse duration to 15 fs. Focusing these pulses into a 1-mm-thick LGS crystal for IPDFG generates waveforms with a spectral coverage from 6 to 11 μm at the −10 dB level and an average power of 54 mW. For gating at 1030 nm, 500 mW average power of the Herriott-cell output are split off by a Fresnel reflection at a fused silica surface.

Alternative gate pulses at 1550 nm were provided by an erbium-fiber laser, broadened and self-compressed in a highly nonlinear fiber (HNLF) [31].[6] The resulting pulses at 100 MHz repetition rate have a FWHM pulse duration of 10 fs and an average power of 165 mW. The results of a second-harmonic generation FROG [10] measurement of the compressed gate pulses are shown in Fig. 4.19.

The temporal intensity envelope (panel (a)) has a FWHM duration of 11 fs. Panel (b) shows the spectral coverage from 1024 to 1750 nm at the −10 dB level. The spectral phase varies within 2 rad, before strongly deviating from its flat behavior at 1100 nm and below. Fused silica plates are used for pre-compensation of the dispersion introduced by the detection crystal. Together with losses at additional optics,

[5] K. Fritsch developed the laser frontend and NIR compression stages. T. Buberl set up the IPDFG, 1-μm EOS and developed the MIR spectrometer with 1030-nm gate pulses (Detailed description in [30])

[6] A. Weigel and P. Jacob developed the 1550-nm EOS. I worked with them on the optimization of the signal strength, investigating different detection crystal thicknesses, the beam geometry and recombination.

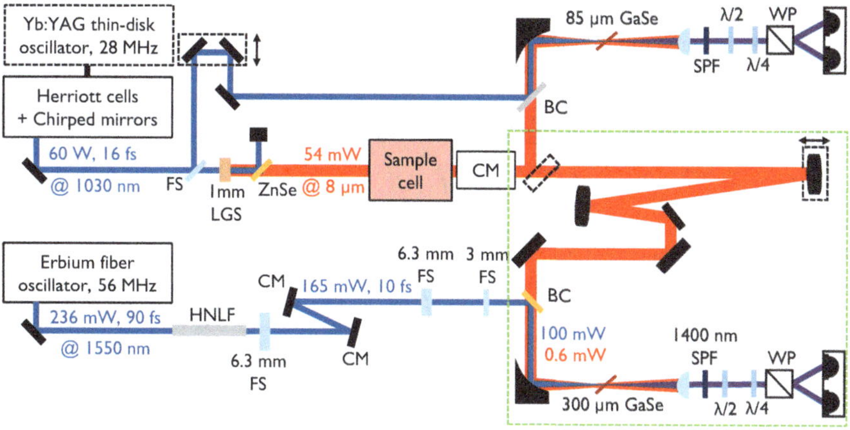

Fig. 4.18 Experimental setup for the comparison of gate pulses at 1030 nm and 1550 nm central wavelength. An Yb:YAG thin-disk oscillator serves as the laser front-end at 1030 nm central wavelength. The pulses are temporally compressed to 16 fs with 60 W of average power in two Herriott-cell-type spectral broadening stages and with chirped mirrors (CM). A small portion of the resulting pulses is split off *via* reflection off a fused silica substrate to serve as the 1030-nm EOS gate pulses. The remaining power generates 8-μm-pulses *via* IPDFG in LGS, which are sent through a sample cell and temporally compressed with chirped mirrors. They can then be optionally directed to EOS with 1030-nm, or 1550-nm gate pulses with a flip mirror. For the 1550-nm gate pulses, the 236-mW, 90-fs output of an erbium-fiber oscillator is compressed to 10 fs in a highly nonlinear fiber (HNLF) and with chirped mirrors. Fused silica is used for pre-chirping of the gate pulse. In the detection setup using 1030-nm gate pulses, the MIR and gate beams are spatially combined with a germanium plate. They are focused into a 85-μm-thick GaSe crystal. A SPF at 950 nm is used for dynamic range optimization. For electro-optic detection at 1550 nm, the beam combiner (BC) is a ZnSe polarizer. A 300-μm-thick Gase crystal is used for upconversion and a 1400-nm SPF for dynamic range optimization. In both detection setups, two waveplates and a Wollaston prism serve as the ellipsometry setup before balanced detection

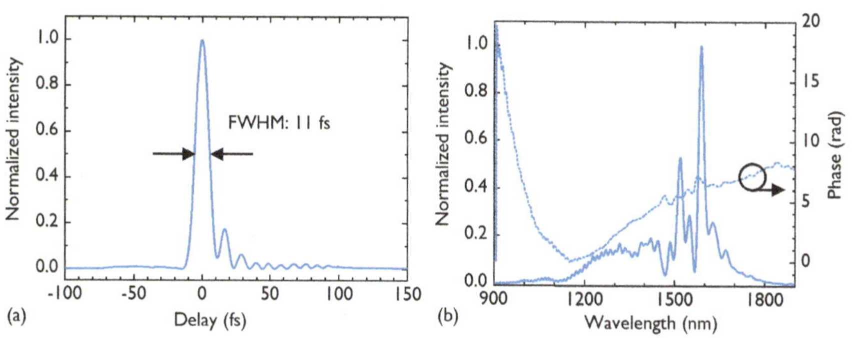

Fig. 4.19 Results of a FROG measurement of the 1550-nm gate pulses. **a** Retrieved temporal intensity envelope with a FWHM pulse duration of 11 fs. **b** Retrieved spectrum and spectral phase

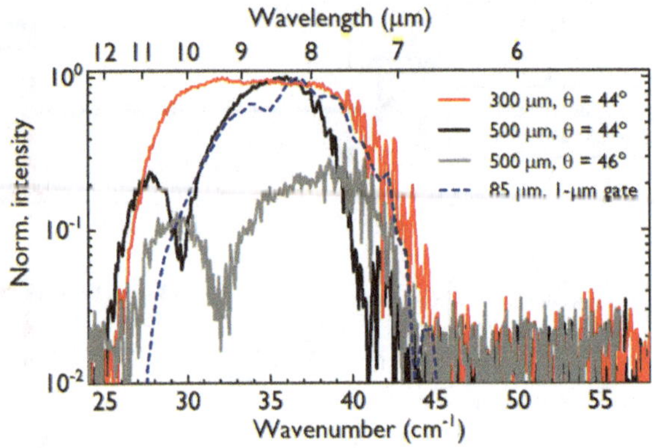

Fig. 4.20 Comparison of EOS spectra measured with the 1550-nm gate pulses for different EOS crystal thickness. The blue dashed line shows the spectral coverage of the 1030-nm EOS with a 85-μm-thick detection crystal, scaled in intensity for comparability

this results in an average gate pulse power of 100 mW at the GaSe crystal. In order to record EOS traces, the second laser was synchronized to the Yb:YAG thin-disk oscillator with a slight offset in repetition rate, thereby implementing asynchronous optical sampling (ASOPS) [12]. The frequency difference leads to a constant delay slip between the two lasers from one pulse to the next, so that a recurring delay scan is performed.

Before recombination with the gate pulses, the MIR field is sent through a sample cell and is temporally re-compressed with chirped mirrors. With the help of a flip mirror, it is directed either to the 1-μm or 1.55-μm gate-pulse detection setup. In the first case, a germanium plate spatially combines the MIR and gate beams before they are collinearly focused onto a 85-μm-thick GaSe crystal for EOS detection. A 950-nm SPF is used for dynamic range improvement. In the latter, a telescope, consisting of two concave mirrors, enlarges the MIR beam diameter by a factor of three, to achieve a focus spot size similar to that of the NIR beam at the position of the nonlinear crystal. A ZnSe polarizer spatially recombines the two beams, before they are focused into the GaSe crystal for electro-optic detection.

As the first step of the electro-optic signal optimization for the 1550-nm gate pulse, the thickness of the GaSe crystal was chosen such that the spectral bandwidth of the signal matches that of the 1030-nm EOS. This allows for a direct comparison of the performance of the different gate pulses, with respect to the spectral coverage, dynamic range, stability etc. Figure 4.20 shows a comparison of the spectrum measured with the 1030-nm gate pulses and a 85-μm-thick detection crystal to several spectra measured with the 1550-nm gate pulses for different crystal thicknesses and phasematching angles.

All spectra show sharp spectral modulations around 7 μm wavelength, caused by water absorption in ambient air. Slight frequency-shifts of the spectra are caused by remaining jitter in the delay axis which will be improved in future experiments. The spectrum measured with a 85-μm-thick crystal using the 1030-nm gate pulses spans from 7.1 to 10.3 μm at −10 dB. Using a 500-μm-thick crystal with the 1550-nm gate pulses results in a similar spectral coverage, however with a distinct dip caused by phasematching. Two different settings of the GaSe rotation angle show the tunability of the spectral response with this parameter. For the 300-μm-thick crystal, the EOS response matches that of the 1030-nm detection at short wavelengths and extends up to 12 μm. We chose this crystal to achieve a flat and broadband spectral response.

In the second step of the signal maximization, the pre-chirp of the gate pulses was optimized for the 300-μm-thick detection crystal. Fused silica is one possible material, as it has a GDD with the opposite sign to GaSe for wavelengths longer than 1200 nm. Different thicknesses were tested to achieve maximum signal strength,[7] with the optimum found for a 9-mm-thick plate, matching the simulation results shown in Sect. 3.5. As the broadband gate pulse spectrum spans from 1024 to 1750 nm at −10 dB (see Fig. 4.19b), but the dispersion relation of fused silica and GaSe is different, perfect pre-chirping with fused silica can only be achieved at one specific gate pulse wavelength. Furthermore, fused silica has its zero-dispersion wavelength at $\approx 1.27\,\mu$m, thus changing the sign of the group velocity dispersion. As shown in the simulations, dispersive mirrors were designed for optimized pre-chirping. Those were implemented at the time of writing this thesis and showed the expected signal increase.[8]

As the next optimization parameter, the MIR beam was expanded with a mirror telescope (see Fig. 4.21a) by a factor of three to match the focal spot sizes of MIR and gate beam at the position of the detection crystal. A manual stage is used to adjust the collimation of the MIR beam with the second concave mirror and thus match the lateral focus position of the two beams

Figure 4.22 shows the matched beam waists at the EOS crystal position, measured with a scanning-slit beam profiler (Nanoscan 2s Pyro/9/5).

The matching focus sizes for both beams are $2w_0 = 44\,\mu$m with a Rayleigh length of $2z_R = 2$ mm for the NIR beam and $2w_0 = 22\,\mu$m with a NIR-Rayleigh length of $2z_R = 490\,\mu$m, respectively. For the smaller focal length, the Rayleigh length of the NIR beam is still longer than the crystal length. Despite the stronger divergence of the MIR beam, which is much smaller than the 300 μm crystal length, we observed a higher signal with stronger focusing. One possible reason is the increased peak intensities of both beams for this configuration.

A KRS-5 and a ZnSe polarizer (both from Thorlabs) were considered for spatial combination of the MIR and gate beams. In comparison to using an uncoated germanium plate, this leads to a power increase at the detection crystal of more than a factor of two in both beams. KRS-5 is beneficial due to its five times lower GDD as compared to ZnSe. However, the NIR beam profile was strongly distorted upon reflection (see the comparison of panels (c) and (d) in Fig. 4.21), thus affecting the EOS signal strength. The final recombination geometry is depicted in Fig. 4.21b, with the MIR beam transmitted through and the gate beam reflected off the ZnSe polarizer.

[7] P. Jacob optimized the pre-chirp.

[8] V. Pervak provided the chirped mirrors, P. Jacob and A. Weigel implemented them in the experimental setup.

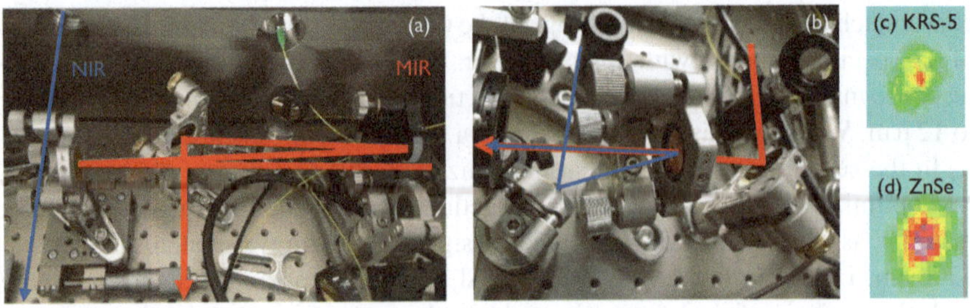

Fig. 4.21 Pictures of parts of the experimental setup, including the MIR beam path indicated with the red and the gate pulse/NIR beam path with the blue line. **a** MIR telescope for focus size adjustment. The beam passes two concave mirrors with focal lengths $f_1 = 75$ mm and $f_2 = 250$ mm and a flat, uncoated gold mirror. The left concave mirror is placed on a translation stage for adjustment of the collimation. **b** ZnSe polarizer, used for recombination of the MIR and NIR beams. The MIR beam is reflected off a flat, uncoated gold mirror before being transmitted through the polarizer. The NIR beam is reflected off a protected silver mirror before being reflected by the polarizer under a small angle. **c** Gate-pulse beam profile behind EOS focus when using a KRS-5 polarizer as a recombiner. (d) Gate-pulse beam profile behind EOS focus when using a ZnSe polarizer as a recombiner

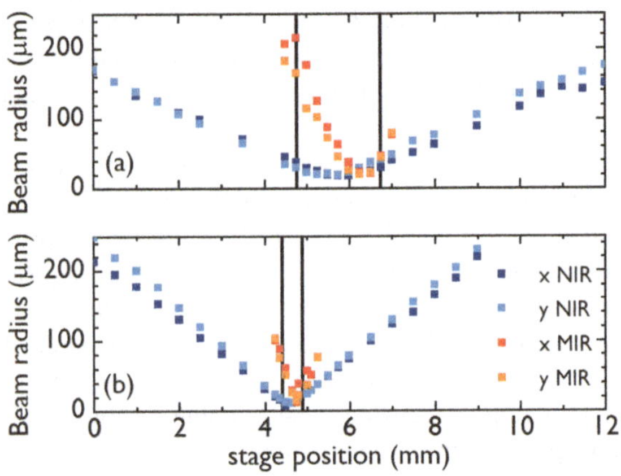

Fig. 4.22 Modematching of the MIR and the 1550-nm gate beam at the position of the EOS crystal. **a** Caustic of the MIR and NIR beams through the EOS focus for a 50.8-mm focal length parabola. The focus diameter for both beams is $2w_0 = 44\,\mu$m. The vertical lines indicate the Rayleigh length of the NIR beam, $2z_R = 2$ mm. **b** MIR and NIR caustic through the EOS focus for a 25.4-mm focal length parabola. The focus diameter for both beams is $2w_0 = 22\,\mu$m with a Rayleigh length of the NIR of $2z_R = 490\,\mu$m

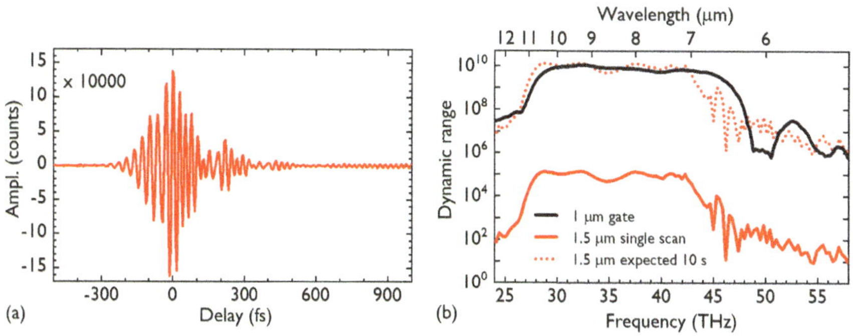

Fig. 4.23 Time-domain trace and frequency-domain intensity dynamic range for the 1550-nm gate pulse EOS. **a** Single-scan time-domain trace with an acquisition time of 100 μs. **b** Single-scan frequency-domain intensity dynamic range (red solid line). Expected dynamic range for a measurement time of 10 s (red dotted line) and comparison to the dynamic range of the slow scan, 1030-nm gate pulse EOS for the same measurement time (black solid line)

The performance of the 1550-nm detections setup in terms of dynamic range under optimized conditions was investigated next. Using a 1400-nm SPF resulted in the highest dynamic range. A single time-domain trace, recorded with the 1550-nm gate pulses in a measurement time of 100 μs, is shown in Fig. 4.23a.

The corresponding frequency-domain intensity dynamic range is shown as the red line in Fig. 4.23b, reaching 1.3×10^5 at 8 μm wavelength. For the 1030-nm gate-pulses, the highest dynamic range achieved with the 85-μm-thick crystal in a typical measurement time of 10 s is shown as the black line in Fig. 4.23b. The red dotted line indicates the expected level of the dynamic range for the 1550-nm detection setup when using the same measurement time, assuming that averaging over that time scale decreases the noise floor with the characteristic $\propto \sqrt{T_{\text{meas}}}$ scaling with the measurement time.

In conclusion, the dynamic range expected for the 1550-nm gate pulses is comparable to what we achieved with 1-μm gate pulses in the same measurement time. Because of better phasematching, a similar spectral coverage was demonstrated with a more than 3 times thicker detection crystal. In addition, the shorter gate pulses would allow for the detection of shorter MIR wavelengths. The optimizations achieved in this work benefit the use of ultra-rapid dual-oscillator EOS for field-sensitive measurements with high sensitivity. Ultra-rapid scanning allows to record full EOS traces faster than most common noise contributions and to investigate fast processes. In the meanwhile, this was implemented in the experiment, also employing the as-level stability of MIR waveforms generated via IPDFG, demonstrated in the previous section, for tracking the timing between the two combs electro-optically [22]. Because of the reduced noise within one trace acquisition, the fast-scanning approach promises increased referencing capabilities and reproducibility for future applications in biomedical spectroscopy [32].

4.7 EOS with 2-μm Gate Pulses

With the thorough optimization and characterization of EOS with 1-μm gate pulses described in Sect. 4.1, the upconversion efficiency in EOS was improved by more than one order of magnitude compared to the state of the art [1]. Nevertheless, due to phasematching and power limitations, the maximum MIR to NIR conversion efficiency was 2%, resulting in an overall maximum detection efficiency of 0.76%. As discussed in Sect. 3.4, the recent development of longer-wavelength laser frontends, based on thulium, holmium and chromium doped active media, allows to simultaneously improve both these parameters. On the one hand, as shown with simulations in Sect. 3.4 and for the erbium-fiber-based gate pulses in the previous section, the electro-optic detection bandwidth for MIR waves employing GaSe improves with increasing gate pulse wavelength. On the other hand, the lower photon energy at longer gate pulse wavelengths increases the damage threshold and thus allows for the use of higher gate pulse powers, which also increases the signal strength.

In this section, proof-of-principle EOS measurements with high-power, 2-μm gate pulses are presented. In comparison to the erbium-fiber-based gate pulses, these allow to exploit also the higher damage threshold of the detection crystal and thus investigate limitations of the detection efficiency, which follows in the next section [33, 34].

The laser system is described in detail in [35], showing the configuration for EOS that was used for the proof of principle experiments. A schematic of this experimental setup is depicted in Figs. 4.24a, b.[9]

An erbium-fiber oscillator from Menlo Systems is amplified and Raman shifted to 1965 nm in a nonlinear fiber. After temporal stretching with a fiber Bragg grating, these pulses seed a thulium fiber amplifier [36], generating a 50-MHz repetition rate pulse train at 1965 nm central wavelength, with 50 W of average power and 250 fs FWHM pulse duration. This beam is split into two channels: In one channel, 7 W of average power are sent into a photonic crystal fiber (PCF) to generate pulses with 13 fs FWHM pulse duration and 4.5 W average power *via* nonlinear self-compression. These pulses are transmitted through a 2-mm-thick sapphire and a 2-mm-thick fused silica plate to compensate for the dispersion of the remaining transmissive optics. A $\lambda/2$ plate and a polarizer are used for attenuation and to ensure linear polarization. The pulses are temporally delayed with a mechanical stage and serve as the gate pulses for electro-optic detection of the MIR waves generated in the second channel with an average power of 500 mW. In the second channel, pulses with 33 W of average power are nonlinearly self-compressed in a second PCF to 32 fs with 30 W of average power. These pulses drive IPDFG in a 1-mm-thick GaSe crystal. Up to 500 mW of MIR power, centered at 8 μm wavelength can be generated with this configuration

---

[9] The laser frontend was developed in a collaboration with the group of Prof. Jens Limpert at the FSU Jena. The fiber compression and MIR generation were implemented by T. P. Butler, D. Gerz and J. Gessner with support from J. Xu and me. As indicated by the green dashed line, I planned and set up the EOS detection, with support from T. P. Butler and D. Gerz. Together with D. Gerz, I optimized the EOS detection.

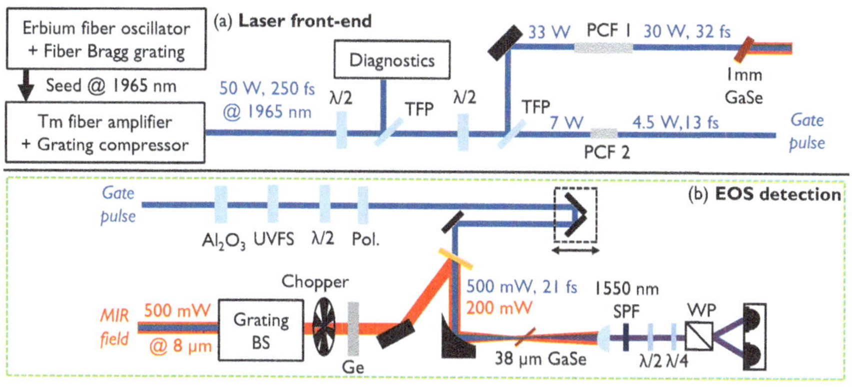

Fig. 4.24 Sketch of the laser system used for EOS with 2-μm gate pulses. **a** Laser front-end: Er-fiber oscillator, amplified and Raman shifted to 1965 nm, temporally stretched with a fiber Bragg grating and amplified in two Tm-fiber amplifiers. The output is sent into two photonic crystal fibers (PCFs) for nonlinear self-compression. Pulses with 30 W of average power and 32 fs FWHM pulse duration from the first fiber generate MIR in a 1-mm-thick GaSe crystal *via* IPDFG. The 4.5-W, 13-fs output pulses from the second PCF serve as gate pulses for EOS. Gold-coated gratings split the MIR and driving pulses and a mechanical chopper modulates the MIR beam for lock-in detection. **b** Beam path for proof-of principle EOS measurements, with the full setup in vacuum. The IPDFG field is transmitted through 6-mm-thick germanium for dispersion compensation of the generation crystal. The gate pulse is pre-chirped with 2-mm-thick sapphire and 2-mm-thick fused silica. A $\lambda/2$ plate and a polarizer are used for attenuation and to ensure linear polarization. The gate pulses are temporally delayed with a mechanical stage. MIR and gate beam are recombined with a ZnSe polarizer. A 38-μm-thick GaSe crystal and a 1550-nm SPF are used for EOS detection

[35, 37]. Gold coated gratings suppress the near-infrared driving pulse power by three orders of magnitude while reflecting more than 80% of the MIR power[10] [38]. Transmission through a chopper wheel with a chopping frequency of 7 kHz serves as the modulation for Lock-in detection of the EOS signal and reduces the average MIR power by a factor of two. A 6-mm-thick, uncoated germanium plate compensates for the dispersion of the IPDFG crystal.[11] The gate and MIR beams are recombined with a ZnSe polarizer, with the second propagating in reflection, resulting in 200 mW of MIR average power arriving at the detection crystal. For characterization of the generated field as it exits the IPDFG crystal, the whole setup was kept at a mbar-level vacuum environment to avoid absorption losses in air and the number of transmissive optics for the IPDFG beam was reduced to a minimum.

In order to infer the electric field of the input waveform from an EOS measurement, the instrument response has to be taken into account, as discussed in Sect. 3.3. The two inputs to this calculation are the detection crystal geometry and the gate pulse properties. In the geometry shown in Fig. 4.24b, the gate pulse is transmitted through several dispersive substrates and polarization optics. We characterized its

[10] The grating beam splitters were implemented by D. Gerz.

[11] T. P. Butler optimized the MIR field compression, see details in [35]

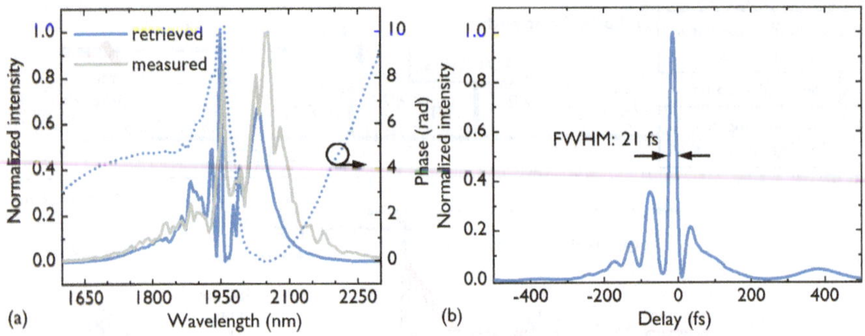

Fig. 4.25 FROG measurement of the gate pulse used for the first 2-µm EOS measurements. **a** Retrieved spectrum and spectral phase (blue), as well as measured spectrum (grey) for comparison. **b** Retrieved time-domain intensity envelope, with as FWHM pulse duration of 21 fs

spectrum and spectral phase with a home-built FROG [10] setup. The results of this measurement are shown in Fig. 4.25.

Panel (a) shows the retrieved spectrum and spectral phase and a measured reference spectrum. The agreement of the retrieved and measured spectra serves as a validation for the FROG measurement. Differences, particularly at longer wavelengths, can be caused by a strong dependence of the shape of the reference spectrum on the spectrometer input coupling. For wavelengths longer than 1950 nm, a strong curvature of the spectral phase is observed. The corresponding retrieved temporal intensity envelope is illustrated in panel (b), with a FWHM pulse duration of 21 fs and 32% of the power contained in the main peak. For the proof of principle experiments with 2-µm gate pulses, we used a 39-µm-thick GaSe crystal rotated to $\theta = 44°$ and $\varphi = 0°$ for EOS detection. A 1550-nm SPF was used to suppress spectral components of the gate pulse spectrum that do not interact with the upconverted photons [1]. With these parameters and the gate pulse FROG measurement, the EOS IRF was calculated numerically as explained in Sect. 3.3—see the results in Fig. 4.26.

The frequency-domain response in panel (a) has a three-peak structure, with a FWHM spectral width spanning from 6 µm to 10 µm and a long-wavelength wing extending up to 16 µm. In the time domain (see panel (b)), this corresponds to a convolution of a pulse with 26 fs intensity-FWHM duration, resulting from the gate pulse duration and dispersion in the detection crystal.

The electro-optic signal was measured as the difference of the currents from two InGaAs diodes (Hamamatsu, G1218-005), amplified with a transimpedance amplifier (DLPCA-200 from FEMTO Messtechnik GmbH). The resulting voltage was detected with a lock-in amplifier (Zurich Instruments, UHFLI) with a time constant of 166 µs, at the modulation frequency of the chopper of 7 kHz. The measured EOS trace is shown as the green line in Fig. 4.27a, the corresponding intensity spectrum in panel (b).

The intensity FWHM of the measured trace is 43 fs, with the spectrum spanning from 8 µm to 16 µm at the −10 dB level. With the calculated instrument response

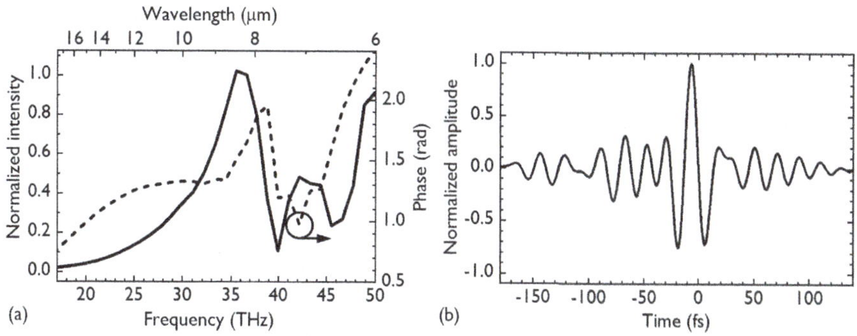

Fig. 4.26 Simulated instrument response of EOS with 2-μm gate pulses for a 39-μm-thick, GaSe detection crystal, rotated to $\theta = 44°$ and $\varphi = 0°$, and using a 1550-nm SPF for noise reduction of the local oscillator. **a** Intensity and phase of the frequency-domain IRF. **b** Normalized amplitude of the time-domain IRF

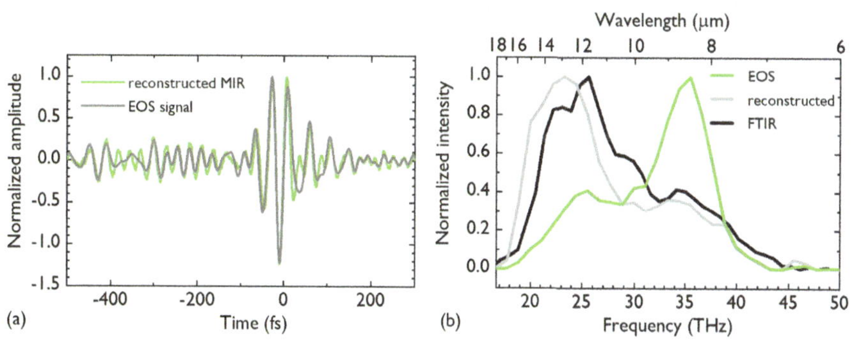

Fig. 4.27 Measured and retrieved field and spectrum for the first EOS with 2-μm gate pulses. **a** Measured EOS trace (green) and retrieved MIR field (grey) for a 39-μm-thick EOS crystal. **b** EOS spectrum (green), retrieved spectrum (grey) and measured FTIR spectrum (black). Figure modified from [35]

from Fig. 4.26, the electric field was retrieved from the measured EOS trace—see the grey line in Fig. 4.27a. The agreement of the reconstructed MIR and the measured FTIR spectrum, plotted in Fig. 4.27b, confirms the validity of the calculated IRF.

4.8 Optimized 2-μm EOS: Reaching Percent-Level Detection Efficiency

After proof-of-principle EOS measurements with 2-μm gate pulses for the characterization of the generated IPDFG waveforms, the electro-optic detection was optimized to maximize the photon detection efficiency and dynamic range, similar to the steps performed for the 1-μm gate pulses described in Sect. 4.1.

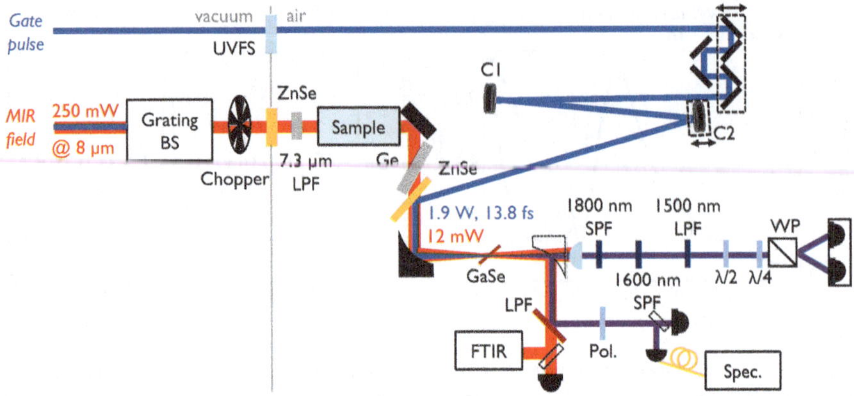

Fig. 4.28 Beam path for detection efficiency optimized EOS. A 1-mm-thick ZnSe and a 1.5-mm-thick fused silica valve window (UVFS) allow for operation of the EOS detection under ambient pressure while keeping the IPDFG in vacuum. The 7.3-μm long-pass filter (LPF) suppresses signal from water absorption lines and the sample cell may be used for gas-phase measurements. A 1-cm-thick germanium plate compensates for the ZnSe dispersion. Two concave mirrors (C1, C2) allow for NIR beam size and divergence adjustment. The MIR and NIR beams are spatially combined with a ZnSe plate and focused into a GaSe crystal for EOS detection. After the crystal, the transmitted MIR and generated sum-frequency spectra and powers can be characterized. For EOS detection, a combination of three spectral filters optimizes the local oscillator contributions and two waveplates and a Wollaston prism serve as the ellipsometry setup before balanced detection

Beam Geometries

The first improvement was the reduction of transmissive optics in the gate pulse beam path, resulting in the adapted beam geometry for electro-optic detection shown schematically in Fig. 4.28 and in pictures in Fig. B.1.

Transmission through the waveplate and two polarizers was avoided by using a wedged, 1-mm-thick ZnSe substrate, placed under Brewster's angle, for recombination of the MIR and NIR beams. The only additional material transmission after the self-compressing fiber was a 1.5-mm-thick fused silica window in the vacuum valve. This adds the same amount of GDD as 250 μm of GaSe at 1870 nm, but with opposite sign. The shorter wavelength spectral wing of the gate pulse spectrum is thus pre-chirped for a 500-μm-thick detection crystal. Pre-chirping of the gate pulse is discussed in more detail for the 1550-nm gate pulses in Sect. 3.5. The results of a FROG measurement of the optimized gate pulse at the position of the electro-optic detection crystal are shown in Fig. 4.29.

The temporal intensity envelope shown in panel (a) has a FWHM duration of 13.8 fs, with 56% of the power contained in the main peak. Panel (b) shows the retrieved spectrum and spectral phase and an independently measured spectrum for comparison and validation. The reduced material transmission along with the Fresnel reflections also increased the gate pulse power arriving at the detection crystal by almost a factor of four, to 1.9 W. Furthermore, the EOS focus sizes and positions along the propagation direction of the MIR and gate beams were matched by

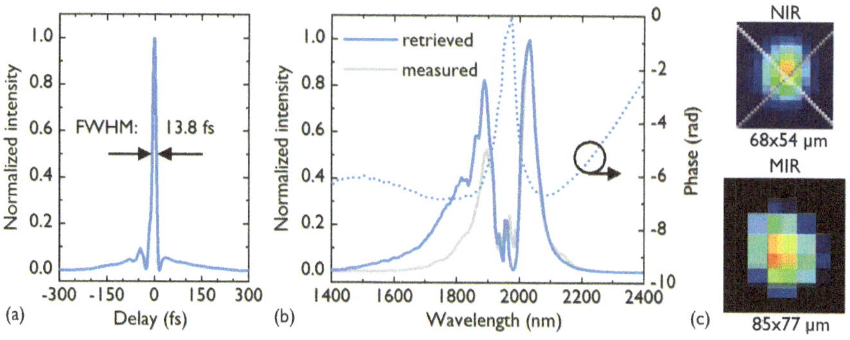

Fig. 4.29 FROG measurement of the optimized gate pulse and NIR and MIR beam profiles at the EOS crystal. **a** Retrieved time-domain intensity envelope, with a FWHM pulse duration of 13.8 fs. **b** Retrieved frequency-domain spectrum and spectral phase, and independently measured spectrum for validation. **c** Measured MIR and NIR beam profiles at the position of the EOS crystal, with the $1/e^2$ beam sizes given for two orthogonal axes

adjusting the NIR size and divergence with a telescope consisting of two concave mirrors. The resulting beam profiles are shown in Fig. 4.29c and were measured using a WinCamD-UHR using two-photon absorption and a WinCamD-FIR2-16-HR for the NIR and MIR beams, respectively. According to Eq. (2.37), their average $1/e^2$ diameters of $2w_{\mathrm{gate}} = 61\,\mu\mathrm{m}$ and $2w_{\mathrm{MIR}} = 81\,\mu\mathrm{m}$, result in total spatial overlap of $> 90\%$. The MIR beam is transmitted through a 1-mm-thick ZnSe window in a vacuum valve, such that the full beam-path can either be kept under vacuum environment or the EOS detection can be optimized in air. Furthermore, a sample cell with with two wedged, 1-mm-thick ZnSe windows may be used for gas-phase absorption measurements (see Chap. 5). Temporal compression of the MIR field is achieved with a 1-cm-thick germanium substrate, placed under Brewster's angle for minimum losses. Furthermore, a long-pass filter (LPF, coated 1-mm-thick germanium) with the cut-on wavelength at 7.3 μm can be placed in the beam to avoid unwanted background signals from water absorption in gas-phase measurements.

Comparison of Different Detection Crystal Thicknesses

As described in Sect. 2.4 in a simplified picture, which does not include dispersion and spatial or temporal walk-off of the interacting pulses, the power conversion from the MIR to the NIR in sum-frequency generation scales with the square of the crystal thickness. However, for thick crystals, these effects lead to deviations from this scaling law, limiting the maximum achievable power conversion. For investigating the maximum MIR photon detection efficiency in EOS in the present experimental configuration, several crystal thicknesses were tested and compared in terms of MIR to NIR conversion efficiency, signal strength, spectral coverage and balanced noise floor. The overall MIR photon detection efficiency was then characterized for the crystal with the best trade-off in terms of detection efficiency and bandwidth.

Table 4.1 Comparison of nominal and measured detection crystal thicknesses

Nominal thickness (μm)	Reflection position (ps)	Thickness (μm)
750	13.6	736
500	10.0	541
300	5.7	308
100	2.4	130

Five different crystals were compared, nominally 100, 300, 500, 750-μm and 1-mm-thick. Due to uncertainties in the manufacturing process, the nominal and actual crystal thicknesses can differ significantly. For comparison of the data sets, we determined the actual thickness for each case from the delay position of the first inner reflection of the infrared beam in the EOS trace. The MIR travelled twice through the crystal with thickness d and group refractive index $n_{\mathrm{MIR}} \approx 2.7$ at the central wavelength of 9 μm, resulting in the following relation for the thickness and the delay of the inner reflection τ_{MIR}:

$$d = \frac{c\tau_{\mathrm{MIR}} \cos(\theta_{\mathrm{int}})}{2n_{\mathrm{MIR}}} \tag{4.4}$$

The internal propagation angle $\theta_{\mathrm{int}} \approx 12.9°$ of the MIR beam inside the GaSe crystal increases the effective propagation length. The resulting thickness values are summarized in Table 4.1.

The measured thicknesses deviate from the nominal ones by up to 30%. For the 1-mm-thick crystal, no reflection could be observed, most likely due to spatial separation.

For all measurements presented below, the gate pulse was similar to the FROG measurement shown in Fig. 4.29 and the spatial overlap and detection crystal rotation and position were optimized to maximize the time-domain signal strength. Figure 4.30 shows the measured EOS traces for all crystal thicknesses.

As described in the previous section, the signal amplitude was measured as the amplified difference current from two InGaAs diodes, now detected at a chopping frequency of 9.5 kHz with a lock-in time constant of 166 μs and a filter of order six, corresponding to ≈ 2.2 ms total integration time per data point. For the 541-μm to 1-mm-thick crystals, the MIR power on the detection crystal was ≈ 12 mW. For the thinner specimens, the IPDFG field was compressed with an additional 3 mm of Germanium, in order to reduce the longer wavelength tail of the EOS traces that is visible in upper two traces. Inserting the additional Germanium plate reduced the power to 5.8 mW, the EOS traces were therefore scaled with the square-root of the power ratio. A 2-mm-thick, anti-reflection coated fused silica substrate in the gate beam path was used to introduce additional pre-chirp for the 1-mm-thick crystal. For the two thickest crystals, the local oscillator power had to be attenuated to avoid saturation of the current amplifier. Furthermore, while the gate pulse spectrum and average power were kept approximately constant for all measurements, slight

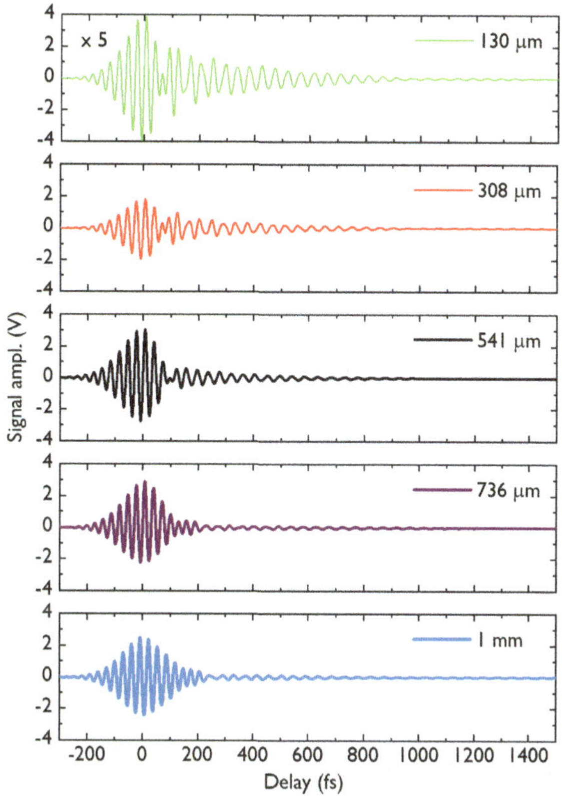

Fig. 4.30 Time-domain electro-optic signal for various detection crystal thicknesses. The signal is measured as the amplified difference current of two diodes. From top to bottom, the detection crystal thickness was increased from 130 μm to 1 mm. All traces were scaled with the square-root of the MIR power on the detection crystal, the square-root of the local oscillator power and the NIR power attenuation factor after the electro-optic detection crystal

deviations in the spectral broadening caused power changes of the local oscillator—that is in the short wavelength wing of the gate spectrum—by more than a factor of four. To allow for comparison of the signal strengths, the traces were scaled with the square-root of the local oscillator power and with the attenuation factors, respectively.

Two effects are observable in the measured traces with increasing detection crystal thickness: Firstly, an increase of the signal strength up to 541 μm and secondly, a reduction of the signal in the longer-wavelength 'tail'. The suppression of the longer wavelengths can be explained by considering the changes of the frequency-domain spectra, i.e., the decrease in phasematching bandwidth, as illustrated in Fig. 4.31.

Panel (a) shows the intensity spectra on a linear scale, as obtained from a Fourier transform of the traces shown in Fig. 4.30. The figure also includes a reference MIR spectrum measured with an FTIR spectrometer (compact lamellar FTIR spectrometer from LASNIX). For better visibility of the data for the thinnest detection crystals, panel (b) shows the same spectra on a logarithmic scale. The signal intensity values in panel (c) summarize the scaling of the signal strength at 9.2 μm (32.5 THz). With increasing thickness, the spectral narrowing that was already visible in the time-domain traces, is observed. Furthermore, the signal strength at 9.2 μm does

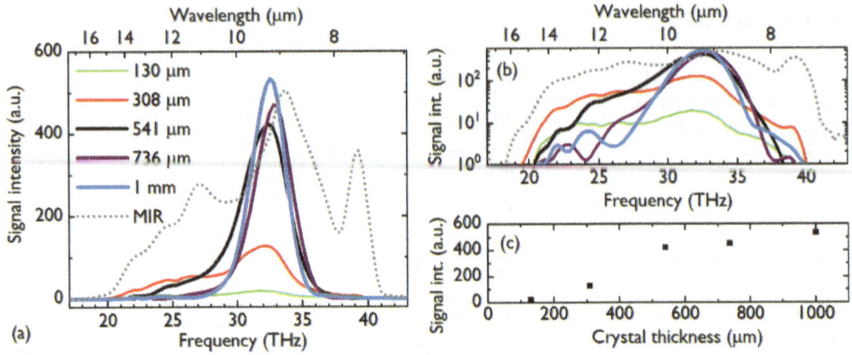

Fig. 4.31 EOS spectra for all crystal thicknesses. **a** Intensity spectra, obtained as the Fourier transform of the traces in Fig. 4.30. The grey dashed line shows the MIR spectrum transmitted through the detection crystal as measured with an FTIR, for comparison. **b** Same spectra as in **a**, but on a logarithmic scale for better visibility of the measurements for the thinnest crystals. **c** Scaling of the signal intensity at 9.2 μm/32.5 THz

not increase significantly for crystals thicker than 541 μm. In the following, the MIR photon conversion efficiency is characterized to investigate the reasons for this signal scaling.

Characterization of the MIR to NIR Conversion Efficiency

Two main factors limit the overall MIR photon detection efficiency: Firstly, the nonlinear conversion efficiency of MIR and gate field into the sum-frequency field. Secondly, sum-frequency photons are lost at the spectral filters used for selection of the local oscillator photons—see calculations in Sect. 2.6.

As the first parameter, the MIR-to-NIR upconversion efficiency was characterized. Therefore, the power of the infrared waveform transmitted through the GaSe crystal was measured as a function of the gate pulse delay with a thermal power sensor (μW power sensor from LASNIX), analogously to the measurements described in Sect. 4.1. The resulting delay-dependent transmission measurements are shown in Fig. 4.32a.

All measurements were normalized to the power transmission outside the temporal overlap region of gate pulse and MIR field. Therefore, power losses due to Fresnel reflection at the crystal or (possible) delay-independent nonlinear effects of the MIR field itself do not contribute to the measurements. With increasing crystal thickness, the minimum power transmission increases from 2% for the 130-μm-thick crystal by more than one order of magnitude to 27.7% for the 1-mm-thick one. The minima of the depletion measurements as a function of the crystal thickness are summarized as the dark blue dots in Fig. 4.32b. Saturation of the depletion is observed from 541 μm onwards, similar to the EOS signal strength. For the two thinnest crystals, the depletion curves oscillate with a period similar to the MIR wavelength. Most likely, these oscillations are caused by an interference of radiation generated inside

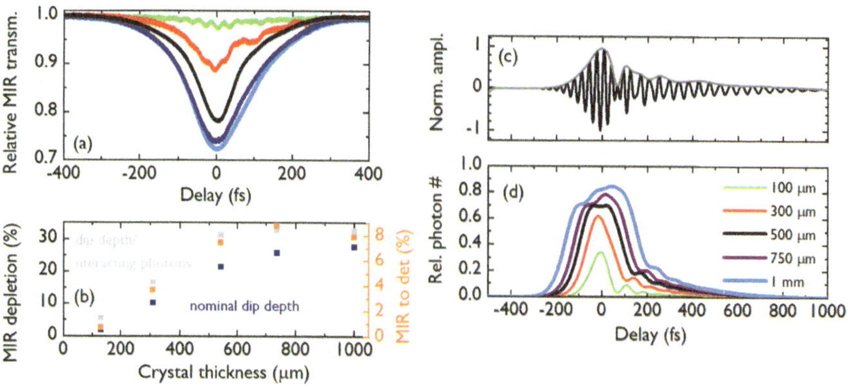

Fig. 4.32 Spectrally integrated and spectrally resolved MIR power depletion. **a** Spectrally integrated MIR depletion as a function of the gate pulse delay for varying crystal thickness. **b** Minimum transmission as a function of the detection crystal thickness, as nominally measured in **a** (dark blue) and corrected for the number of interacting photons, as calculated in **d** (grey). The orange dots on the right axis show the number of mid-infrared photons in the EOS crystal within the interaction time-window detectable after additional losses in the heterodyning setup at the peak of the signal. **c** MIR waveform as measured with a 130-μm-thick detection crystal. **d** Relative number of interacting photons as a function of the gate pulse delay. Calculated from the intensity envelope of the MIR field in **c** and the gate-pulse walk-off and dispersion

the detection crystals and the transmitted field or the delay-dependent interference of the sum-and difference frequency components (see Sect. 3.1).

Because the temporal walk-off between MIR and gate pulse as well as dispersion increase with the crystal thickness, the effective interaction time window of the two fields elongates. Therefore, and depending on the shape of the characterized waveform, the relative number of MIR photons interacting with the gate pulse varies with the crystal thickness and the gate pulse delay. For a normalization of the spectrally integrated dip measurement to the number of interacting photons, the interaction time-window was calculated for each thickness, considering dispersion and temporal walk-off of the gate pulses. The effective interaction window was then defined as spanning from the time corresponding to half the intensity maximum at the onset of the gate pulse at the beginning of the crystal, to half the intensity value at the falling edge of the dispersed intensity envelope at the end of the detection crystal.

From the thinnest to the thickest detection crystal, the temporal FWHM of the interaction window increases from 58 to 367 fs. The shape of the MIR field was assumed to closely follow the trace measured with the thinnest crystal, as the green line in Fig. 4.31c is similar to the spectrum measured with the FTIR. For reference, this measurement is shown again in Fig. 4.32c. Knowing the shape of the field incident on the detection crystal allows for the calculation of the number of MIR photons within the interaction time window for each delay point, as shown in Fig. 4.32d. The maximum percentage of interacting photons increases from 35% for the 130-μm-thick crystal to 85% for the 1-mm-thick one. The maximum dip depth related to the number of interacting photons is plotted as the grey dots in Fig. 4.32b. The observed saturation is even stronger than the non-normalized one, with maximum values around 30%, indicating no further increase of the nonlinear interaction. This relative depletion is assumed to be constant for all gate-pulse delays, with deviations caused by the chirp of the MIR waveform.

Overall, we measured a maximum average MIR photon conversion efficiency reaching 30%. Saturation of the depletion is most likely caused by temporal walk-off and dispersion in the thickest detection crystals. The temporal walk-off limits the interaction-time of the gate pulse at a delay position with the largest number of MIR photons, the second reduces the peak intensity of the gate pulse and therefore lowers the interaction strength. A further increase of the conversion efficiency could thus only be reached with even higher gate pulse powers, up to the point of back-conversion and saturation due to a limited number of available MIR photons available for interaction (see also the simulation results in Sect. 3.2).

The measurement of the overall power depletion does not reveal potentially significantly stronger depletion at certain wavelengths. The percent-level depletion and the mW-level MIR powers available in the experiment allowed for spectrally resolved measurements with an FTIR spectrometer (see the setup sketch in Fig. 4.28). For all detection crystal thicknesses, FTIR spectra of the MIR beam transmitted through the detection crystal were recorded while scanning the gate pulse delay. Figure 4.33 shows the spectrally resolved relative intensity changes, obtained from the measured spectra by normalization to a reference spectrum, which was recorded for a gate pulse delay away from the temporal overlap with the MIR field.

In Fig. 4.34, the spectral depletion is plotted for the gate pulse delay of maximum electro-optic signal.

The spectrally resolved measurements show depletion over almost the full spectrum for the two thinnest crystals, however with a maximum of 25% for the 308-μm-thick crystal. With increasing thickness, the depleted spectral region narrows down and increases in depth, reaching down to only 40% transmission at 33 THz for the 1-mm-thick crystal. Starting from the 541-μm-thick crystal, the MIR depletion saturates, similar to the observations from the spectrally integrated measurements. Furthermore, for the 1-mm-thick crystal, the optimum spatial overlap differs for maximum depletion and maximum EOS signal, as illustrated with the depletion measurements in Fig. 4.35.

Rotating the GaSe crystal by 2° to a flatter phasematching angle, thus tuning the spectral response to longer wavelengths, and adjusting the spatial overlap of MIR and gate beam with one mirror in the MIR beam path, allowed us to locally reach a spectral depletion of up to 80%. However, angular walk-off in the thick detection crystal causes the spatial overlaps for maximized signal and maximized depletion to differ significantly. Furthermore, changing the alignment of the MIR beam effectively changes the propagation angle inside the detection crystal, which can explain the changing wavelength of best depletion for the two different alignments. A test with even thicker crystals (namely 2 mm) confirmed that this behavior becomes even more severe. We therefore concluded that in addition to the saturation of the depletion, spatial walk-off limits the maximum achievable signal strength and the optimum

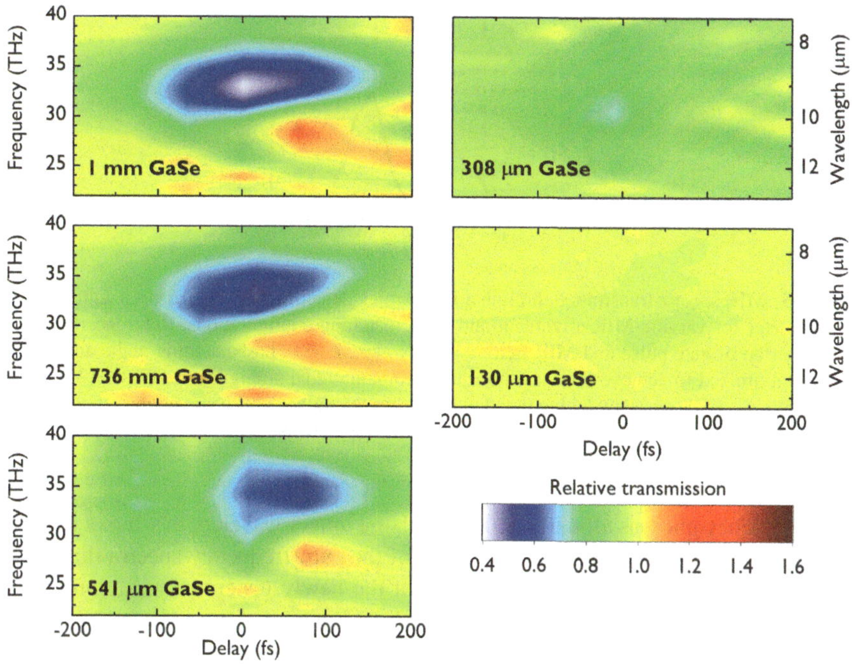

Fig. 4.33 Delay-dependent spectrally resolved MIR depletion for several detection crystal thicknesses, obtained by normalization of the measured gate-pulse delay-dependent spectra transmitted through the detection crystal to a reference spectrum acquired outside the temporal overlap region

Fig. 4.34 Spectrally resolved depletion at the gate pulse delay of maximum electro-optic signal strength for several detection crystal thicknesses

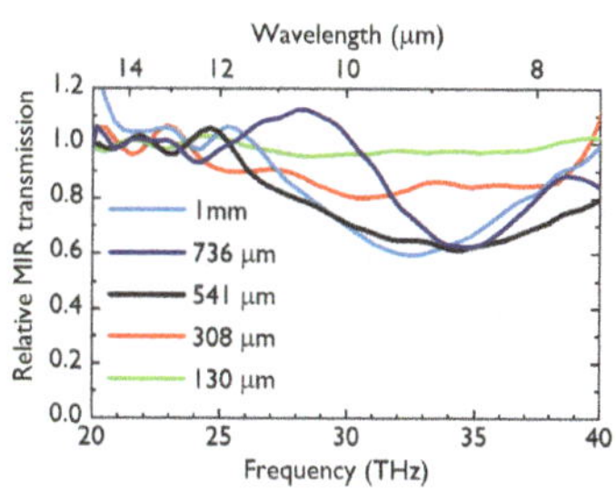

crystal would be $\approx$ 500 μm thick, where both effects are not yet dominating. Furthermore, the detection bandwidth to signal strength trade-off is best for this thickness. The effects of spatial walk-off could be avoided when using crystals which can be cut to a certain phasematching angle and therefore be placed at 0° angle of incidence.

In Figs. 4.33 and 4.34, significant generation of MIR radiation in the detection crystal is visible. After the minimum, the three thickest crystals show delay-dependent, local increases at $\approx$ 28 THz of up to 60%. As this effect is delay dependent, we assume the MIR generation to be a cascaded effect, appearing only as a consequence of the interacting pulses. One possibility is DFG of the newly generated

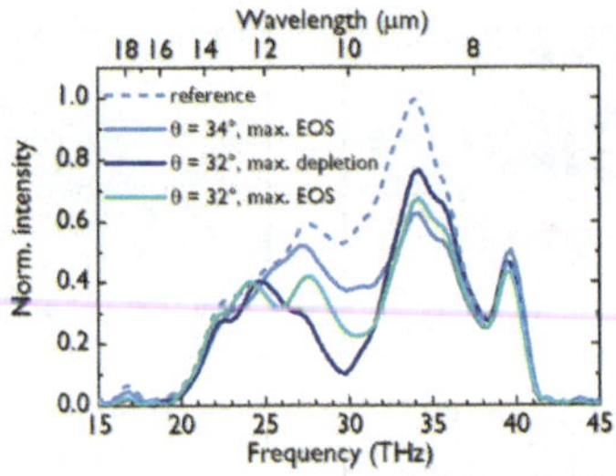

Fig. 4.35 MIR spectra transmitted through a 1-mm-thick detection crystal measured with an FTIR spectrometer for varying MIR alignment and phasematching configurations. Reference: No temporal overlap of gate pulse and MIR field. $\theta = 34°$, max. EOS: Phasematching angle and spatial overlap optimized to achieve the maximum time-domain signal strength in EOS. $\theta = 32°$, max. depletion/max. EOS: Phasematching tuned to longer wavelengths, MIR beam alignment to achieve optimum depletion or maximum EOS signal strength, respectively

sum-frequency photons around 1700 nm with the longer wavelength wing of the gate pulse. The delay-dependence matches that of the sum-frequency spectra illustrated in Fig. 4.36 and explained below. Effects of the newly generated radiation on the linearity of the EOS detection will be investigated in Sect. 4.9.

Generated Sum-Frequency Spectra and Spectral Filtering

The second step of EOS is heterodyne detection of the generated sum-frequency photons with the transmitted gate pulse. We measured the generated spectra as a function of the gate pulse delay behind a wire-grid polarizer rotated to transmit p-polarized light, thus suppressing the local oscillator spectrum, see Fig. 4.36.

The spectra are normalized to the measured sum-frequency power, which is consistent with the number of depleted MIR photons. Two dashed lines indicate the spectral window from 1500 to 1600 nm, which is selected with the filter combination described below. For all crystal thicknesses, sum-frequency components are generated in this spectral range at delays before the EOS signal maximum. After the peak, almost all the generated components have wavelengths longer than 1600 nm. For the two thinnest crystals, a long-lasting modulation at even longer sum-frequency wavelengths is observed. The modulation itself can be explained by interference of sum and difference frequency components, as both processes can occur with similar strengths for thin crystals (see Fig. 3.1 for the simulation results for a 1-μm-thick crystal). In addition, where those modulations stretch into the spectrally filtered region, the long-wavelength tail of the EOS traces is measured (see Fig. 4.30). The delay-dependent change of the wavelength of maximum sum-frequency signal can be explained by the spectral phase of the gate pulses. This is illustrated by the simulation results shown in Fig. 4.37.

Panel (a) shows the simulated sum-frequency spectra and EOS signal for a Fourier-transform limited gate pulse. For panel (b), the gate pulse was taken from the FROG measurement shown in Fig. 4.29, therefore pre-chirped for a 500-μm thick detection crystal using fused silica. Most likely, a combination of chirp and temporal walk-off

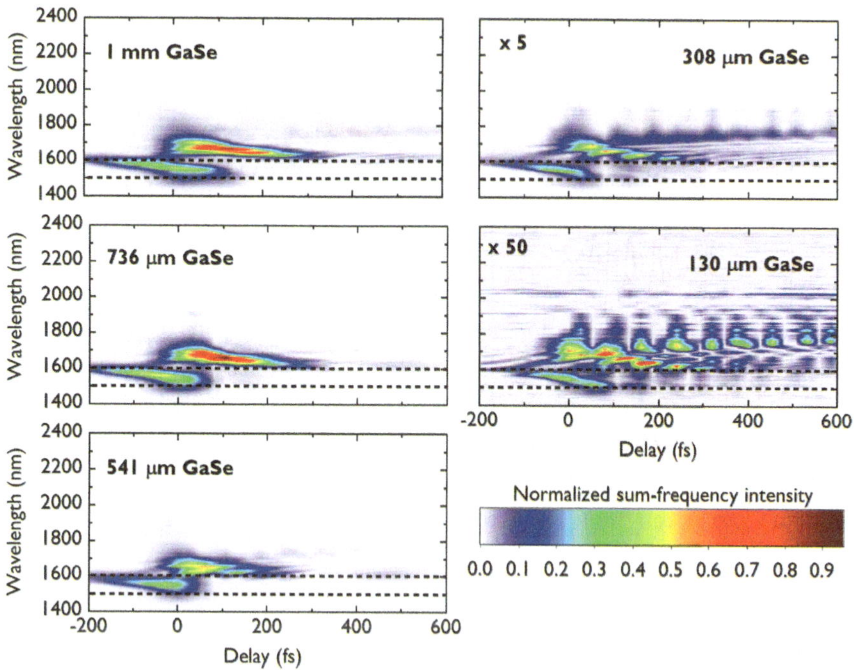

Fig. 4.36 Generated sum-frequency spectra as a function of the gate pulse delay for several detection crystal thicknesses. The dashed lines indicate the spectral region transmitted to the detector

results in different mixing process depending on the relative temporal position of gate pulse and MIR field at the beginning of the nonlinear interaction. At the same time, the pre-chirp increases the signal strength as it results in an almost frequency-independent EOS signal around the peak of the simulated trace. Further improvement of the pre-chirp with custom optics, as shown for the 1550-nm gate pulses in the previous section, is expected to result in even better signal strengths.

Fig. 4.38a shows the spectrum of the gate pulse and of the sum-frequency spectrum at the maximum of the time-domain EOS signal for a 541-μm-thick detection crystal.

Around 1550 nm, the generated sum-frequency power is approximately six times stronger than the local oscillator power. Because the electro-optic signal is not normalized to the total power, this does not cause nonlinearities within the trace, as will be shown in Sect. 4.9. According to the calculations in Sect. 2.6, the SPF wavelength to achieve maximum dynamic range for a shot-noise limited balanced noise floor is at $\approx$ 1750 nm. However, we obtained the best results with filters resulting in the transmission plotted on the right axis of Fig. 4.38a (a 1500 nm long-pass filter (LPF), a 1600 nm SPF and a 1800 nm SPF[12] to avoid transmission of wavelengths $>$ 2000 nm where the SPF reflection decreases again). Slight differences in transmission are caused by the perpendicular polarizations and the dip in the transmission

[12] Custom coating number IC1802 by V. Pervak.

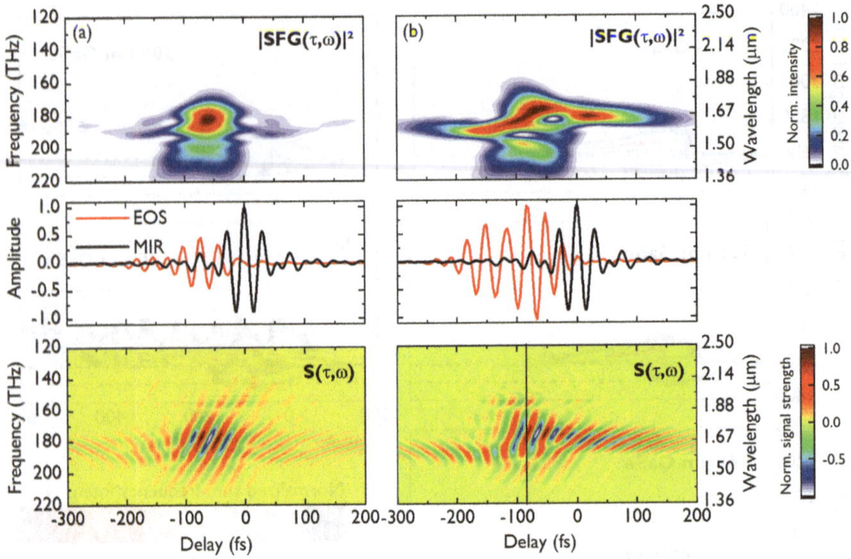

Fig. 4.37 Simulated delay-dependent sum-frequency spectra and wavelength dependent EOS signal for a 500-μm thick GaSe crystal and two different spectral phases of the 2-μm gate pulses. **a** Fourier-transform limited gate pulse. **b** Gate pulse as taken from FROG measurement, pre-chirped with fused silica for the 500-μm thick detection crystal

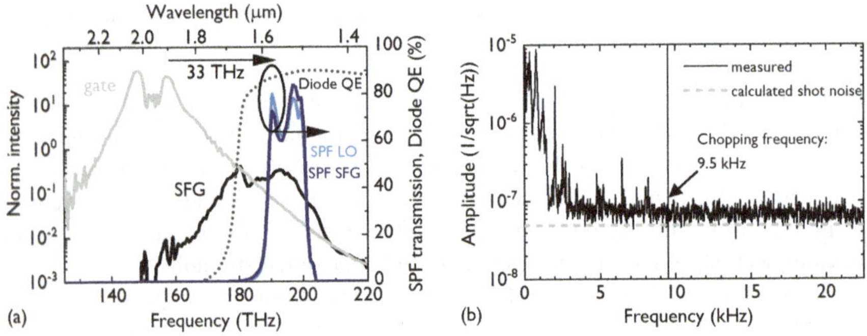

Fig. 4.38 Spectral filtering and balanced noise floor. **a** Gate pulse and maximum sum-frequency spectrum for a 541-μm-thick detection crystal. The right axis shows the measured transmission properties of the final spectral filtering geometry for the local oscillator and sum-frequency. The quantum efficiency of the balanced diodes is plotted as the dashed line on the right axis. **b** Radio-frequency spectrum of the balanced local oscillator noise for the 541-μm-thick EOS crystal. Dashed grey line: Calculated shot noise for 203 μW of local oscillator power per diode. Vertical line: Chopping frequency of 9.5 kHz

curve is caused by the coating of the 1600-nm SPF. The right axis also includes the quantum efficiency of the balanced diodes.[13] In the selected region, the generated sum-frequency at the delay of maximum signal is strongest in comparison to the local oscillator. The arrow indicating a spectral shift of 33 THz indicates that this corresponds to mixing of the short wavelength wing of the gate pulse spectrum with the wavelengths of highest power in the MIR spectrum. On the one hand, filtering at shorter wavelengths decreases the spectral response for long MIR wavelengths, as discussed in Fig. 3.7b and in [39]. Even though the beam sizes on the balanced diodes were maximized and a large bias-voltage was used, as suggested in [40], including longer local oscillator wavelengths decreased the signal to noise ratio, as balancing to the shot-noise limit becomes challenging for broader spectra and higher powers, due to imperfections of the phase retarders and RIN. Therefore, the optimum filter cut-off here was found to be at shorter wavelengths than predicted for shot-noise limited balancing with the calculations shown in Fig. 2.7b. In addition, the strong dip in the center of the gate pulse spectrum limits the wavelengths which mix efficiently and can therefore also limit the detection bandwidth. This effect is to be investigated in future simulations and experiments with a less-modulated spectrum.

Figure 4.38b shows the balanced noise floor after optimization of the power distribution with a combination of a $\lambda/2$ and a $\lambda/4$ plate. The total local oscillator power measured behind the spectral filters was 0.55 mW. Additional losses of optical power at the Wollaston prism, transport optics and the detector quantum efficiency, result in an effective optical power per diode of 203 μW. For the central wavelength of the detected spectrum of 1550 nm, the theoretical shot noise was calculated from Eq. (2.47) and is plotted as the dashed grey line. At the chopping frequency of 9.5 kHz (indicated by the vertical line in Fig. 4.38b), the average balanced amplitude noise does not show significant deviations from a white-noise floor. However, we determined the average noise value to be a factor of 1.4 above the calculated shot noise level. For all other crystal thicknesses, a similar balanced noise floor was measured, thus no worsening of the balancing, caused by crystal birefringence was observed.

While optimizing the detection dynamic range, spectral filtering reduces the number of detectable sum-frequency photons. The number of photons transmitted through the final filter geometry was calculated for each crystal thickness for the peak of the signal. Panel (b) in Fig. 4.32 shows the resulting overall number of MIR photons that arrive inside the detection crystal within the interaction time window, which can contribute to the electro-optic signal after nonlinear conversion and spectral filtering on the right axis. It also includes Fresnel losses of the sum-frequency beam at the back surface of the detection crystal, which could be avoided with suitable coating techniques (see the discussion in Sect. 4.4). The total number of detectable photons varies between 1 and 8.8% for the different thicknesses and reaches 7.5% for the 541-μm-thick detection crystal.

[13] Quantum efficiency of Hamamatsu G12180 calculated from responsivity.

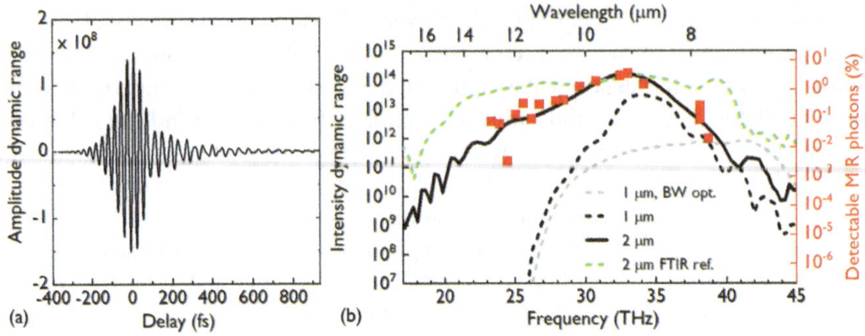

Fig. 4.39 Time and frequency domain dynamic range. **a** Time-domain amplitude dynamic range, measured with a 541-μm-thick crystal with 1 s measurement time per temporal element. **b** Frequency-domain intensity dynamic range, achieved with a 541-μm-thick detection crystal in 16 s total measurement time for 12 mW of average MIR power. Dashed black line: Optimum frequency-domain intensity dynamic range for 1-μm gate pulses, for the same measurement time, average MIR power and nominal detection crystal thickness. Dashed grey line: Dynamic range for 1-μm gate pulses in the broadband configuration. Dashed green line: Spectrum measured with an FTIR spectrometer for the 2-μm experiments. Red/right axis: wavelength dependent percentage of MIR photons, contributing to the electro-optic signal after upconversion and transmission through the post-interaction spectral filtering for the 2-μm EOS

Dynamic Range and Overall Detection Efficiency

Measuring the MIR depletion and the sum-frequency transmission through the SPF for the temporally compressed field allows for an estimation of the spectrally averaged MIR photon number that can be upconverted and sent to the detector for the dynamic range optimized measurement geometry. A direct measure for the detected photons, that includes interference of the local oscillator and sum-frequency photons, the spatial overlap and additional transmission losses, is the time-domain amplitude dynamic range for 1 s measurement time for each temporal element, as shown in Fig. 4.39a.

The time-domain dynamic range, with a maximum of 1.5×10^8, was obtained by dividing the maximum time-domain signal of a single EOS trace (5.04×10^6 V) by the average noise floor for 1 s average time (0.0335 V). For a MIR average power of 12 mW reaching the detection crystal, a flux of $\approx 3.8 \times 10^{17}$ photons per second inside the detection crystal within the interaction time window was calculated using the MIR spectrum shown in Fig. 4.31a (grey dashed line). The maximum average number of detected photons is then given by the square of the ratio of the peak dynamic range to the interacting MIR photons, namely $\approx$ 5.9%. This value is a factor of 1.17 lower than the detectable MIR photons after upconversion and spectral filtering, which is reasonable considering imperfect overlaps, interference and additional losses.

Investigating the spectrally resolved depletion measurements in Fig. 4.33, the photon detection efficiency at different wavelengths can differ significantly. For measuring the MIR-wavelength resolved detection efficiency, the MIR field was temporally

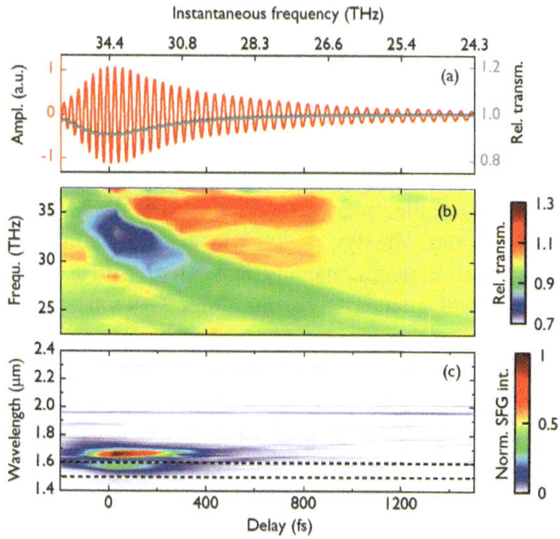

Fig. 4.40 Measurement of the wavelength dependent overall detection efficiency with a chirped MIR waveform. **a** Chirped waveform (red, left y-axis) and spectrally integrated power depletion measurement (grey, right axis). **b** Delay dependent spectrally resolved MIR relative transmission, measured with an FTIR spectrometer. **c** Delay dependent sum-frequency spectra, with the dashed lines indicating the transmission window of the spectral filters

chirped using a 4-mm-thick ZnSe substrate. The resulting waveform, now temporally stretched over more than 1 ps, is shown in Fig. 4.40a.

Panel (a) also shows the spectrally integrated power depletion measurement. The minimum here is reduced by a factor of ≈ 3, as compared to the compressed pulse, because there are less photons in the interaction time window. Panel (b) shows the delay dependent relative MIR depletion measured with the FTIR spectrometer. The relative transmission reaches a minimum of 70% at the instantaneous wavelength of 9 μm. New frequency components around 35 THz (8.6 μm) and 30 THz (10 μm) are generated at delays after the peak of the EOS trace. Panel (c) shows the generated sum-frequency spectra as a function of the gate pulse delay, with the transmission window of the spectral filters indicated by the dashed lines. The maximum transmission of 13.2% is reached at the instantaneous frequency of 33 THz and decreases to $\approx 3\%$ towards longer wavelengths. Most likely, the mid-infrared generation for longer delays is caused by cascaded processes with the long-wavelength sum-frequency radiation (> 1.6 μm), as they show a similar delay-dependence. Combining the depletion and SPF-transmission measurements, the wavelength dependent number of MIR photons that contributes to the electro-optic signal can be calculated as the product of the FTIR dip depth and the transmission through the SPF, resulting in the red dots plotted on the right axis of Fig. 4.39b. At the peak of the spectral response at ≈ 9 μm, the number of detectable photons surpasses $> 3\%$. The steep drop in efficiency is caused by the increasing phase mismatch. Furthermore, detection of longer wavelengths is suppressed by spectral filtering after the nonlinear conversion, as illustrated by the comparison of different filters in Fig. 3.7b and described in [39]. Overall lower numbers for the detectable photons as compared to the measurement for the compressed pulse can be explained by the imperfect separation of the different wavelength components and the reduced number of photons in the interaction time window due to the chirp. Nevertheless, this measurement illustrates the wavelength dependence of the detection efficiency caused by phasematching, the gate pulse parameters and spectral filtering.

This detection efficiency, together with the shape of the MIR spectrum results in the wavelength dependent spectral intensity dynamic range shown in Fig. 4.39b, for a total measurement time of 16 s, and 12 mW average MIR power at the crystal. For comparison, the green dashed line shows the MIR spectrum measured with a FTIR spectrometer. The dynamic range peaks around the wavelength of maximum photon detection efficiency, namely at 9.1 μm with a value of 1.58×10^{14}. At −10 dB, the spectral intensity dynamic range spans from 8.2 to 10.6 μm, nicely matching the shape of the wavelength-dependent detection efficiency. The black dashed line shows the optimum intensity dynamic range achieved with the 1-μm gate pulses (see Sect. 4.2 for the detection optimization and characterization), for the same measurement time, average MIR power and nominal detection crystal thickness. In this case, the narrower spectral coverage is caused by the detection bandwidth. On the other hand, for the bandwidth optimized dynamic range in the 1-μm EOS shown as the grey dashed line, the long-wavelength cut-off stems from the generated MIR spectrum. Comparing the peak intensity dynamic ranges sees an increase of approximately one order of magnitude for the 2-μm gate pulses. This can be related to the increase in overall detection efficiency.

4.9 Linearity of Large Depletion EOS Measurements

Commonly, the theoretical description of EOS assumes undepleted fields and either a strong MIR/THz field with a weak gate pulse or a strong gate pulse interacting with a weak MIR field. With those assumptions, the time-domain signal is—in very good approximation—proportional to the infrared field strength and the intensity of the local oscillator. For a strong gate pulse, the interaction strength is independent on the MIR power. Here, as shown in the previous section for the highly efficient electro-optic detection, the MIR power is depleted by tens of percent, thus violating the assumption of no depletion and suggesting a saturation of the signal strength as a function of the gate pulse power.

Therefore, we investigated the validity of the linear scaling laws for EOS with percent-level detection efficiency. This is crucial for applications like gas-phase spectroscopy, where the time-domain signal from molecules with concentrations varying over several orders of magnitude should ideally be detected with a field-strength independent IRF.

First, we measured the scaling of the EOS signal strength with the gate pulse power. Attenuation of the 13.8-fs-short pulse has to be achieved without introducing dispersion or altering the polarization state, in order to keep the electro-optic response constant in all parameters but the gate power. One option is to use free-standing, 100-nm-thick silicon nitride pellicles, placed at a small angle to avoid changes to the gate pulse polarization. Their refractive index of $n = 2$ at 1965 nm causes Fresnel losses of

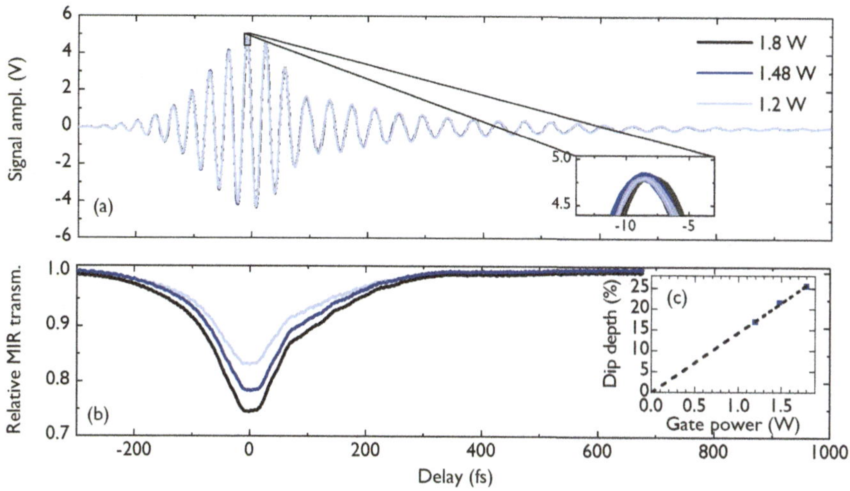

Fig. 4.41 Scaling of the measured EOS signal strength with the gate pulse power. **a** Measured EOS traces for varied gate pulse power, normalized to that power for comparison. The zoom-in to one of the trace extrema illustrates the variations of the normalized signal strength. **b** Spectrally integrated MIR power depletion as a function of the gate pulse delay and power. The inset shows the scaling of the dip depth with the NIR power, with the linear fit through the origin confirming the linear dependence

$\approx 23\%$ upon transmission through one pellicle. We used two pellicles, thus allowing for attenuation of the maximum gate pulse power from 1.8 to 1.48 W (one pellicle) and 1.2 W (2 pellicles). Figure 4.41a shows the resulting EOS traces, scaled with the gate pulse power for better comparability.

The zoom-in on one of the trace maxima elucidates the small differences in signal amplitude, with a variation of 0.5%. Panel (b) shows the gate pulse power dependent, spectrally integrated MIR power depletion measurement. The dip depth scaling with the gate pulse power in panel (c) also shows a linear fit through the origin, thus confirming the linear dependence. No saturation of the MIR depletion or signal strength was observed for the W-level gate pulse powers. This behaviour agrees with the simulations shown in Sect. 3.2, which predicted that we are only at the onset of saturation with comparable gate pulse powers.

Secondly, the scaling of the signal strength with the MIR power was investigated. In a first step, the maximum power at the detection crystal of 12 mW was attenuated with a 0.3 and 1.0 OD germanium neutral density filter. This resulted in 6 mW and 1.4 mW reaching the GaSe crystal, respectively. The EOS traces, normalized by the square-root of the power attenuation factor, are shown in Fig. 4.42a.

Zooming in on one of the peaks of the time-domain trace, shows variations of the amplitude by 2.7%. Attenuation by only one order of magnitude still allowed for the MIR depletion measurement. Panel (b) illustrates the measured spectrally integrated power depletion for varied MIR power. For a gate power much larger than the MIR

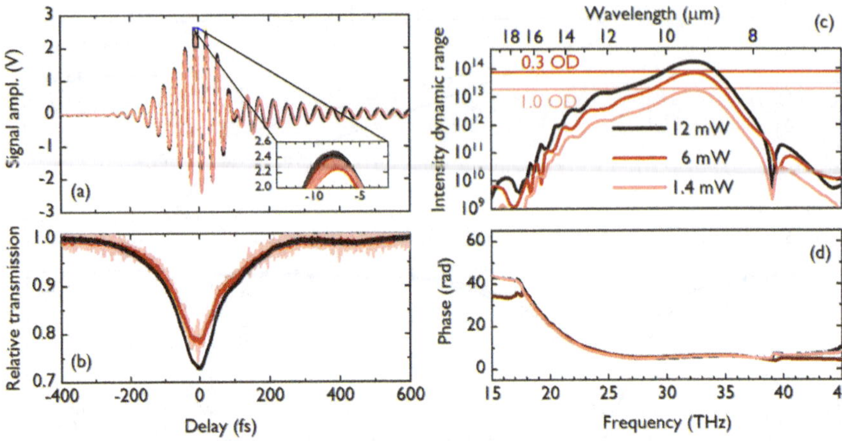

Fig. 4.42 Scaling of large depletion EOS measurements with the MIR field-strength. **a** Measured EOS traces, scaled with the square-root of the power-attenuation factor. The zoom-in on one of the trace extrema illustrates the differences. **b** MIR depletion measurements as function of gate pulse delay for varied infrared powers. **c** Spectral intensity dynamic range. Horizontal lines show expected levels for the indicated MIR power attenuation. **d** Corresponding spectral phases

one, the sum-frequency generation is in the no-pump-depletion regime described in [20]. In this case, the conversion efficiency only depends on the gate pulse power. However, we reproducibly measured 27% maximum depletion for the highest MIR power, as compared to 21% for the other two power values. This increased dip depth matches the increase in signal strength observed in panel (a). The frequency-domain intensity dynamic range for varied MIR power is shown in panel (c), including horizontal lines to indicate the dynamic range expected from the power attenuation at 9.3 μm wavelength. Similarly to the time-domain data, the dynamic range for the highest MIR power is slightly higher than expected, but the spectral phases in panel (d) show no clearly visible deviations. Possible explanations for the more efficient depletion are the contribution of generated sum-frequency radiation to the MIR depletion and an observable influence of the high MIR power on the strength of the nonlinear interaction.

For investigating the linearity of the electro-optic signal over the full measurement dynamic range, an attenuation series of the MIR power over 14 orders of magnitude was performed. A combination of ZnSe and germanium neutral density filters was used, keeping the overall dispersion as constant as possible with possible variations in Fig. 4.43a.

The measurement time for no attenuation up to 13.3 OD was 10 s, averaging over 10 scans, each 2 ps long and with an integration time per temporal element of 2.2 ms. The resulting noise floor is plotted as the dashed black line. For the largest attenuation of 14.3 OD we averaged over twice the number of scans, thus doubling the measurement time and reducing the noise floor, which is on the level of the x-axis. A constant offset in the scaling is visible for the lowest attenuation factors, however

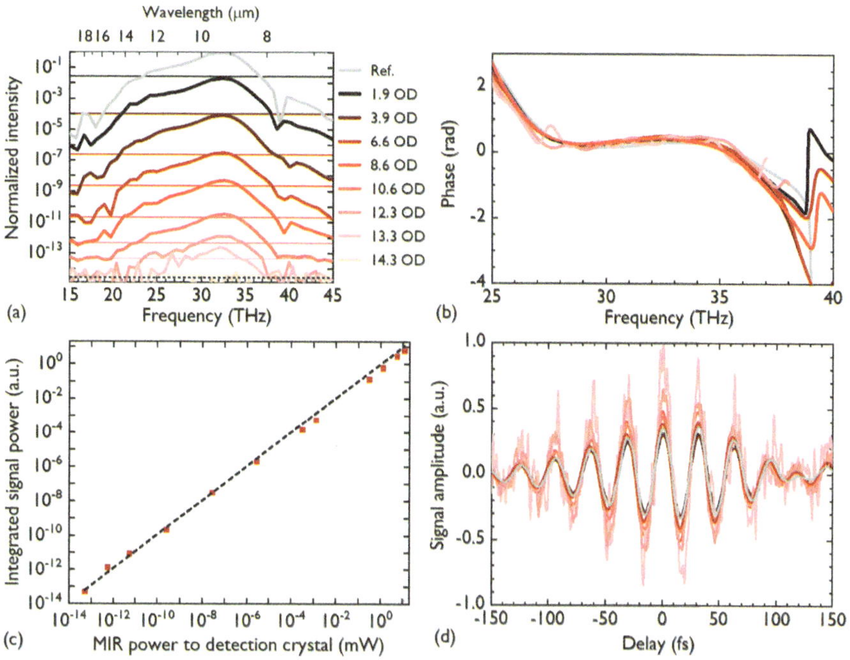

Fig. 4.43 MIR power attenuation series. The individual traces covered a delay-range of 2 ps, with 2.2 ms integration time per temporal element and a measurement time per trace of 1 s. **a** Intensity spectra, not normalized. Horizontal lines indicate the expected signal levels after multiplication of the reference measurement with the nominal attenuation at 9.3 μm. The x-axis is at the noise-floor for the 14.3 OD measurement and the dashed black line shows the noise floor for all other measurements. **b** Spectral phases for the same attenuation factors as the spectra in panel **a**. **c** Spectrally integrated signal power as a function of the nominal MIR power arriving at the detection crystal after attenuation. The attenuated power was calculated from the filter specifications at 9.3 μm. **d** EOS traces, normalized by the square-root of the nominal power attenuation factor at the central wavelength of 9.3 μm

no large deviations from the linear behaviour are observable, which is also confirmed by the similar spectral phases plotted in panel (b). Panel (d) shows the time-domain traces, normalized by the square-root of the nominal attenuation at 9.3 μm. In panel (c), the spectrally integrated signal intensity for the attenuation factors in Figs. 4.42b and 4.43a is plotted as a function of the nominal MIR power reaching the detection crystal, including a linear fit through the origin. Discrepancies between expected and measured values can be caused by imperfect filters and changes in spatial alignment, with some of the filter substrates being slightly wedged.

The measured signal for 14.3 OD attenuation confirms our claimed intensity dynamic range for 10 s measurement time. In addition, the average number of sum-frequency photons reaching the detector in the integration time window at the peak of the EOS trace can be estimated: Attenuating the MIR power by 4.95×10^{13} orders of magnitude (the time-domain intensity dynamic range for 2.2 ms average time per temporal element), an average number of ≈ 22 photons arrives inside the detection crystal in the integration time of 2.2 ms per temporal element of an EOS trace, corresponding to 2×10^{-4} photons per pulse, which is similar to the $< 10^{-3}$ photons per pulse cited in [41] for a measurement time of 1 s. This value corresponds to a minimum measurable field strength of $\approx 4\,\mathrm{V/m}$.

In conclusion, proof-of principle EOS measurements with 2-μm gate pulses were demonstrated for characterization of the high-power infrared fields generated with the same laser system. A thorough optimization of the detection geometry allowed us to employ 1.9 W of gate pulse average power, thus boosting the nonlinear frequency conversion efficiency in the electro-optic detection crystal above 20%. In combination with spectral filters to maximize the detection dynamic range, an average of 6.4% of the MIR photons in the detection crystal and interaction time window arrive at the balanced diodes, requiring only ≈ 22 MIR photons inside the detection crystal in the integration time window per temporal element for the minimum detectable signal. For linear vibrational spectroscopy, this corresponds to a minimum measurable concentration, which is only a factor of 4 above what would be possible when detecting all MIR photons in the interaction time window. Remarkably, within our measurement accuracy, the electro-optic signal strength scaled linearly with the MIR field strength for photon numbers varying between $\approx 10^3$ and 10^{17} per second, despite the large depletion. This means, that enough MIR photons are available for nonlinear interaction with the gate, despite the large depletion and that the MIR power does not significantly determine the interaction strength. An onset of nonlinearity was observed for the highest MIR power, where the interaction strength seemed to increase. Further investigations of the linearity would require an increase of the MIR intensity on the interaction crystal.

References

1. Porer M, Ménard J-M, Huber R (2014) Shot noise reduced terahertz detection via spectrally postfiltered electro-optic sampling. Opt Lett 39:2435–2438. ISSN: 1539-4794
2. Pupeza I et al (2020) Field-resolved infrared spectroscopy of biological systems. Nature 577:52–59. ISSN: 1476-4687
3. Hussain SA (2021) Field resolving spectrometer for mid-infrared molecular spectroscopy PhD thesis, Ludwig-Maximilians-Universität München. https://edoc.ub.uni-muenchen.de/27620/7/Hussain_Syed_Ali.pdf
4. Fritsch K, Poetzlberger M, Pervak V, Brons J, Pronin O (2018) All-solid-state multipass spectral broadening to sub-20 fs. Opt Lett 43:4643–4646. ISSN: 1539-4794
5. Weitenberg J et al (2017) Multi-pass-cell-based nonlinear pulse compression to 115 fs at 7.5 μJ pulse energy and 300 W average power. Opt Exp 25:20502–20510
6. Huber M et al (2017) Active intensity noise suppression for a broadband mid-infrared laser source. Opt Exp 25:22499–22509
7. Pupeza I et al (2015) High-power sub-two-cycle mid-infrared pulses at 100 MHz repetition rate. Nat Photon 9:721–724. ISSN: 1749-4885
8. Amotchkina T et al (2019) Broadband dispersive Ge/YbF3 mirrors for mid-infrared spectral range. Opt Lett 44:5210–5213. ISSN: 1539-4794

9. Schweinberger W et al (2019) Interferometric delay tracking for low-noise Mach-Zehndertype scanning measurements. Opt Exp 27:4789–4798
10. Trebino R et al (1997) Measuring ultrashort laser pulses in the time-frequency domain using frequency-resolved optical gating. Rev Sci Instrum 68:3277–3295. ISSN: 0034-6748
11. Tomasino A et al (2013) Wideband THz time domain spectroscopy based on optical rectification and electro-optic sampling. Sci Rep 3:3116
12. Janke C, Först M, Nagel M, Kurz H, Bartels A (2005) Asynchronous optical sampling for high-speed characterization of integrated resonant terahertz sensors. Opt Lett 30:1405. ISSN: 0146-9592
13. Weigel A et al (2021) Ultra-rapid electro-optic sampling of octave-spanning mid-infrared waveforms. Opt Exp 29:20747. ISSN: 1094-4087
14. Redkin RA et al (2019) GaSe crystals with antireflection coatings for terahertz generation. Mater Res Exp 6:126201. ISSN: 2053-1591
15. Bushunov AA et al (2019) Fabrication of anti-reflective microstructures on chalcogenide crystals by femtosecond laser ablation. Opt Mater Exp 9(1689). ISSN: 2159-3930
16. Hagner M et al (2021) Ultrabroadband suppression of mid-infrared reflection losses of a layered semiconductor by nanopatterning with a focused ion beam. Opt Exp 29:33632. ISSN: 1094-4087
17. Knorr M et al (2018) Ultrabroadband etalon-free detection of infrared transients by vander-Waals contacted sub-10-μm GaSe detectors. Opt Exp 26:19059–19066. ISSN: 1094-4087
18. Ahmed T et al (2017) Mid-infrared waveguide array inter-chip coupling using optical quilt packaging. IEEE Photon Technol Lett 29:755–758. ISSN: 1041-1135
19. Hussain SA, Schweinberger W, Buberl T, Hofer C, Pupeza I (2019) Train of ultrashort mid-infrared pulses with sub-mrad carrier-envelope phase stability in 2019 conference on lasers and electro-optics Europe & European quantum electronics conference (CLEO/Europe-EQEC), IEEE, Piscataway
20. Boyd RW (2008) Nonlinear optics, 3rd edn. Academic Press, Burlington
21. Weiner AM (2009) Ultrafast optics. Wiley, Oxford. ISBN: 0471415391
22. Weigel A et al (2021) Attosecond-precision dual-oscillator infrared field-resolved spectroscopy employing electro-optic delay tracking in 2021 conference on lasers and electro-optics Europe & European quantum electronics conference (CLEO/Europe-EQEC). IEEE, vol 1. ISBN:978-1-6654-1876-8
23. Huber R et al (2001) How many-particle interactions develop after ultrafast excitation of an electron-hole plasma. Nature 414: 286–289. ISSN: 1476-4687
24. Schubert O et al (2014) Sub-cycle control of terahertz high-harmonic generation by dynamical Bloch oscillations. Nat Photon 8:119–123. ISSN: 1749-4885
25. Mayer B et al (2014) Sub-cycle slicing of phase-locked and intense mid-infrared transients. New J Phys 16:063033. ISSN: 1367-2630
26. Somma C, Folpini G, Reimann K, Woerner M, Elsaesser T (2016) Phase-resolved two-dimensional terahertz spectroscopy including off-resonant interactions beyond the chi(3) limit. J Chem Phys 144:184202
27. Sulzer P et al (2020) Passive elimination of correlated amplitude fluctuations in ultrabroadband supercontinua from highly nonlinear fibers by three-wave mixing. Opt Lett 45:4714–4717. ISSN: 1539-4794
28. Kübler C, Huber R, Tübel S, Leitenstorfer A (2004) Ultrabroadband detection of multi-terahertz field transients with GaSe electro-optic sensors: approaching the near infrared. Appl Phys Lett 85:3360–3362. ISSN: 0003-6951
29. Kübler C, Huber R, Tübel S, Leitenstorfer A (2005) Ultrabroadband terahertz pulses: generation and field-resolved detection. Semicond Sci Technol 20:S128. ISSN: 0268-1242
30. Buberl T (2021) Towards next-generation molecular fingerprinting, PhD thesis. Ludwig-Maximilians-Universität München
31. Brida D, Krauss G, Sell A, Leitenstorfer A (2014) Ultrabroadband Er: fiber lasers. Laser Photon Rev 8:409–428. ISSN: 18638880

32. Süss B, Ringleb F, Heberle J (2016) New ultrarapid-scanning interferometer for FT-IR spectroscopy with microsecond time-resolution. Rev Sci Instrum 87:063113. ISSN: 1089-7623
33. Hofer C et al (2020) Mid-infrared electric field sampling approaching single-photon sensitivity. EPJ Web Conf 243:16001
34. Hofer C et al (2021) Electro-optic sampling with percent-level detection efficiency in 2021 conference on lasers and electro-optics Europe & European quantum electronics conference (CLEO/Europe-EQEC), IEEE, vol. 1. ISBN: 978-1-6654-1876-8
35. Butler TP et al (2019) Watt-scale 50-MHz source of single-cycle waveform-stable pulses in the molecular fingerprint region. Opt Lett 44:1730–1733. ISSN: 1539-4794
36. Gaida C et al (2018) High-power frequency comb at 2 mu wavelength emitted by a Tm-doped fiber laser system. Opt Lett 43:5178. ISSN: 0146-9592
37. Butler TP et al (2019) Multi-octave spanning, Watt-level ultrafast mid-infrared source. J Phys Photon 1:044006. ISSN: 2515-7647
38. Gerz D et al (2019) Mid-infrared long-pass filter for high-power applications based on grating diffraction. Opt Lett 44:3014–3017. ISSN: 1539-4794
39. Sulzer P et al (2020) Determination of the electric field and its Hilbert transform in femtosecond electro-optic sampling. Phys Rev A 101:033821. ISSN: 1050-2947
40. van der Valk NCJ, Schouten RN, Planken PCM (2005) Noise suppression of a differential detector under high levels of illumination, relevant to terahertz electro-optic sampling. Rev Sci Instrum 76:073104. ISSN: 0034-6748
41. Riek C et al (2015) Direct sampling of electric-field vacuum fluctuations. Science (New York, N.Y.) 350:420–423. ISSN: 1095-9203

Chapter 5
Conclusion and Outlook

The aim of this thesis was the optimization and characterization of the MIR photon detection efficiency for field-sensitive measurements with EOS. To that end, high gate pulse peak-powers were employed, typically only available at kHz-repetition rate laser systems, which are used for the generation and sensitive detection of THz waves [1]. In addition, phasematching advantages when using longer-wavelength gate pulses allowed the use of thick detection crystals without strict limitations to the spectral coverage. For EOS in the MIR spectral region, established gate pulse sources are erbium-fiber lasers at multi-MHz repetition rates and thin EOS crystals are used to detect waveforms closely matching the electric fields to investigate fundamental quantum optical effects [2–6]. The tight focusing geometries used in those systems result in peak powers comparable to the high-power 2-μm EOS described here, however also limit the interaction length for efficient conversion.

Characterizing the MIR photon detection efficiency in EOS is particularly interesting in two limits: Firstly, for low MIR powers, the photon detection efficiency determines the minimum number of photons required to measure non-zero signal. This is critical e.g., for the detection of weak molecular signatures containing only a few photons. Secondly, for high MIR and gate pulse powers, and the corresponding potentially large conversion efficiency, the fundamental question arises to what extent the linear dependence of the EOS signal on the amplitude of the investigated field and the gate pulse power is a valid assumption.

In this thesis, the MIR photon detection efficiency was first optimized and characterized using 1-μm gate pulses from an Yb-YAG thin-disk oscillator. Measuring the MIR power transmitted through the nonlinear crystal as a function of the gate-pulse delay was introduced for evaluating the MIR to sum-frequency conversion efficiency. This efficiency was limited to 2% by damage of the 500-μm-thick detection crystal with an unprecedented high gate pulse power of 450 mW on a 70-μm-diameter spot at 28 MHz repetition rate and for 16-fs-long pulses. After additional losses in the heterodyne detection and spectral filtering, the overall average detection efficiency

© The Author(s), under exclusive license to Springer Nature Switzerland AG 2022
C. Hofer, *Detection Efficiency and Bandwidth Optimized Electro-Optic Sampling of Mid-Infrared Waves*, Springer Theses,
https://doi.org/10.1007/978-3-031-15328-0_5

was 0.76%. For the 1-μm gate pulses, the trade-off between detection efficiency and bandwidth was critical, caused by the phasematching properties of GaSe. Using a 85-μm thick detection crystal, the full generated MIR spectrum, spanning from 6.6 μm to 10.7 μm at −20 dB was detected, however with two orders of magnitude less detection efficiency. In addition, for the thin detection crystal, multiple reflections of the MIR beam on the surfaces of the detection crystal arise in the tail of the EOS trace. These distort the temporal window in which molecular signals are detected. Several approaches to suppress these reflections by means of optical contacting to materials with matching refractive index were investigated, however with limited success. New developments in other working groups, using micro-structures acting as anti-reflection coatings are promising approaches for future improvements [7, 8]. Nevertheless, the detection efficiency optimization contributed to the finalization of a field-sensitive spectrometer that is currently used for real-world biomedical applications [9] and is outperforming state-of-the-art FTIR spectrometers in sensitivity by up to a factor of 40. In addition, the unprecedented sensitivity and dynamic range allowed for the demonstration of EOS as a broadband and highly sensitive characterization technique for the stability/reproducibility of and, therefore, control over optical fields. Timing fluctuations of the MIR waveforms of < 10 as were measured and the theoretical model developed in the frame of this thesis allowed to conclude that the remaining instabilities can be traced back to the RIN of the laser frontend. The jitter values are ≈ 3 orders of magnitude above of what would be expected from a shot-noise-limited source.

In theory and experiment, the phasematching properties of GaSe were investigated when using longer-wavelength gate pulses. Gate pulses with a central wavelength of 1550 nm are used in our group for the implementation of dual-oscillator scanning to acquire EOS traces faster than most technical noise sources, with the goal to improve the reproducibility and referencing capabilities [10, 11]. Despite the comparably lower average power of the erbium-fiber front-end, a similar performance in terms of detection bandwidth and dynamic range as with the 1-μm system was achieved. This was enabled by the use of thicker detection crystals that boosted the conversion efficiency, while maintaining the detection bandwidth due to better phasematching. With potentially higher gate pulse powers, there is still room for improvement to the detection efficiency using 1550-nm gate pulses. Furthermore, the conversion efficiency is still to be characterized.

For field-resolved spectrometers based on 1-μm sources, one future development in our working group will be the generation and field-resolved detection of MIR wavelengths below 6 μm to spectrally cover absorption lines in the second amide band [12] and thus increase the information obtained from biomedical samples. Generation of short-wavelength radiation via IPDFG requires compression of the driving pulses to < 10 fs pulse duration, which is currently investigated. In addition to pushing the short-wavelength limits of IPDFG in LGS, PPLN [13] is a suitable nonlinear crystal for efficient generation of MIR radiation down to 5 μm. For electro-optic detection, the corresponding un-poled material, namely $LiNbO_3$ is to be investigated as the detection crystal. One challenge when using thick detection crystals will be the gate pulse dispersion, which becomes more critical for ever shorter gate pulses that are necessary to detect the shortest wavelength components.

Secondly, instruments for field-resolved spectroscopy, based on chromium laser-front-ends with a central wavelength at 2.4 μm, are currently under development in our group. They are promising due to their compactness, cost efficiency and the highly efficient MIR generation using IPDFG in ZGP [14, 15], which only requires few-W driving lasers. For developing a field-sensitive spectrometer, electro-optic detection using 2.4-μm gate pulses will be required in those systems. While they are expected to behave similarly to the 2-μm gate pulses presented here, the use of extended InGaAs diodes might not be avoidable. Their noise properties will have to be investigated and potentially the use of cooled diodes will be required.

In addition, new materials like ZGP are to be investigated as EOS detection crystals, promising high detection efficiencies even with moderate gate pulse powers.

During the past years, several high-power laser sources with direct output in the MIR spectral region were developed. One of the most promising technologies is based on thulium-fibers, with a typical central wavelength of 1965 nm. Using such a laser front-end, EOS with 2-μm, 1.9-W gate pulses was performed and optimized for the first time in the scope of this thesis. Several detection crystal thicknesses were then investigated regarding scaling laws of the signal strength, MIR power conversion and the bandwidth versus detection efficiency trade-off. Firstly, the MIR to NIR conversion was pushed to $> 20\%$ using crystals with thicknesses ≥ 541 μm, requiring only a minimum of ≈ 20 MIR photons in the integration time-window. For the measurement of e.g., the free-induction decay in gas-phase absorption spectroscopy, this world-record detection efficiency corresponds to a limit of detection that is only a factor of 4 above what would be achievable when detecting all MIR photons. This sensitivity was meanwhile applied for the detection of ppb-level concentrations of several molecular species.

Significant depletion of the MIR power upon electro-optic detection violates the usual assumption of the nonlinear interaction taking place in the non-depletion regime. Therefore, the scaling of the signal strength with the gate pulse power and MIR field strength was investigated. For attenuation of the gate pulse power from 1.9 W to 1.4 W, no deviation from the linear scaling was observed in the experiment. Simulation results predicted a saturation of the depletion with even higher gate pulse powers, which will be investigated experimentally in the future. For this case, potential advantages of phasematching DFG instead of SFG in EOS were discussed. However, for MIR radiation generated in the detection crystal that is of similar strength as the input beam, the linearity of the signal strength is to be investigated.

Measurements of the EOS signal strength for MIR photon numbers varying between 10^3 and 10^{17} per second, showed no deviations from linear scaling with the electric field strength within our measurement accuracy. However, an onset of nonlinearity was observed for 12 mW of MIR power, with a percent-level stronger depletion than for lower powers. For high gate pulse powers, strong MIR fields and 10%-level depletion of the power of the measured waveform in the detection crystal, the following nonlinear effects should be further investigated:

- For high gate pulse powers and a comparatively low number of MIR photons in the case of large depletion, saturation effects with crystal length and the gate pulse power might occur, due to a limited number of photons available for interaction after depletion.
- Usually, the gate pulse intensity is assumed to determine the strength of the nonlinear interaction. For MIR powers comparable to the gate pulse, both intensities affect the interaction strength. This is a potential source for the deeper dip that we observed in the experiment, but has to be investigated with even higher MIR powers.
- Furthermore, for a constant relative depletion of the infrared field, the number of converted gate pulse photons increases with the MIR power, potentially lowering the detection efficiency for significant depletion before the end of the detection crystal.
- Cascaded nonlinear effects, caused by the generated sum-frequency radiation could distort the electro-optic signal. The polarization of the generated MIR wavelengths that was observed in the experiment is to be investigated in this respect. Regarding the generation of new NIR frequencies, MIR/THz-induced second-harmonic generation is an additional nonlinear effect used for field-characterization [16], that could be observable with our experimental setup.

Gas-Phase Spectroscopy with Percent-Level Detection Efficiency EOS

One application of the high-dynamic range and unprecedented sensitivity of EOS demonstrated with the 2-μm gate pulses is gas-phase absorption spectroscopy. A typical example for the state-of-the-art detection limit in gas phase spectroscopy was shown in [17]: In a multi-pass-cell with a length of 25 m, they detected 9 ppb of ethanol using a quantum cascade laser as the light-source and 10 s measurement time.

Using the formalism presented in [9], we can estimate the limit of detection (LOD) for the intensity dynamic range demonstrated with the 2-μm EOS. Two gases with absorption features in the spectral region of highest detection efficiency are methanol and methane. Methanol has the largest absorption cross section of $\sigma_{\text{methanol}} = 1.4 \times 10^{-18} \frac{\text{cm}^2}{\text{Molecule}}$ at 9.7 μm, where we determined an intensity dynamic range of 8.8×10^{13} in 16 s measurement time, where the 3.3-ps-long scan-window corresponds to $\approx 4\,\text{cm}^{-1}$ spectral resolution. For the planned interaction length of 45 cm, this results in an expected LOD for methanol in an ambient pressure nitrogen environment of $\text{LOD}_{\text{methanol}} \approx 0.15$ ppb. The strongest methane absorption occurs at 7.7 μm, with a cross section of $\sigma_{\text{methane}} = 8 \times 10^{-19} \frac{\text{cm}^2}{\text{Molecule}}$. At this wavelength, the intensity dynamic range is 1.4×10^{12}, thus giving and expected LOD for methane of $\text{LOD}_{\text{methane}} \approx 2$ ppb. Angle tuning of the generation and detection crystals allows to match the spectral region of best sensitivity even better to the absorption frequencies of the molecules.

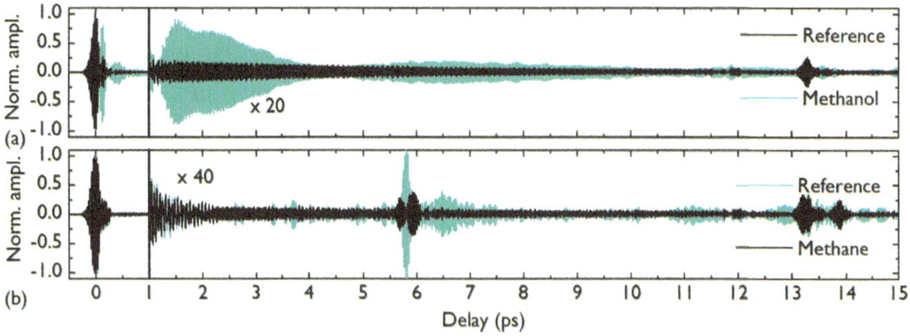

Fig. 5.1 Time-domain EOS traces for proof-of-principle gas-phase absorption measurements of methanol (**a**) and methane (**b**)

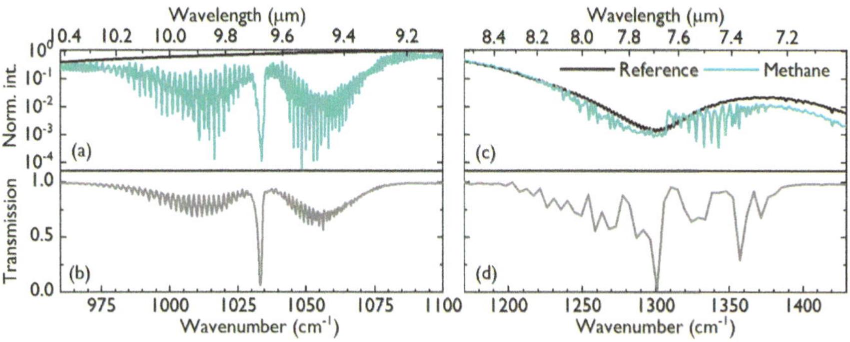

Fig. 5.2 **a** Mesured absorption spectrum of 20 mbar methanol in vacuum environment. **b** Methanol absorption spectrum from [18]. **c** Measured absorption spectrum of methane under ambient air pressure. **d** Methane absorption spectrum from [18]

Here, proof-of-principle measurements with high gas concentrations (20 mbar methanol in vacuum, and methane in air-filled cell) are shown. Exemplary time-domain EOS traces are shown in Fig. 5.1 and the corresponding absorption spectra in Fig. 5.2[1].

The time-domain trace for methanol shows strong distortions of the main pulse due to the large concentration. In the frequency domain, the main features of the reference spectrum (from [18]) are well-reproduced. For methane, no distortion of the main pulse was observed, due to lower concentration and an absorption line away from the peak of the EOS spectrum. Correspondingly, the molecular response in the tail of the pulse is also weaker. The methane absorption lines are in a region of less spectral intensity and where the spectrum is modulated, but are still well-distinguishable. Both time-domain traces show multiple reflections from transmissive optics like cell windows and the generation and detection crystals. All optics except the non-

[1] The methane and methanol samples were prepared by A.-K. Raab, who also helped with the measurements.

linear crystals can be replaced with wedged substrates in the final measurement configuration. Together with an optimized vacuum environment, which suppresses signals from water and other compounds in ambient air, this is crucial to achieve a background-free measurement window.

The verification of the predicted LOD and detailed studies on gas-phase absorption spectroscopy using the 2-μm laser system are subject to current studies. The broad spectral coverage also promises advantages in the detection of several substances with spectrally wide-spread absorption lines [19].

References

1. Krauspe P, Banerji N, Réhault J (2020) Effective detection of weak terahertz pulses in electro-optic sampling at kilohertz repetition rate. JOSA B 37:127–132. ISSN: 1520-8540
2. Kröll J et al (2007) Phase-resolved measurements of stimulated emission in a laser. Nature 449:698–701. ISSN: 1476-4687
3. Huber R et al (2001) How many-particle interactions develop after ultrafast excitation of an electron-hole plasma. Nature 414:286–289. ISSN: 1476-4687
4. Schubert O et al (2014) Sub-cycle control of terahertz high-harmonic generation by dynamical Bloch oscillations. Nat Photon 8:119–123. ISSN: 1749-4885
5. Riek C et al (2015) Direct sampling of electric-field vacuum fluctuations. Science 350:420–423. ISSN: 1095-9203
6. Benea-Chelmus I-C, Settembrini FF, Scalari G, Faist J (2019) Electric field correlation measurements on the electromagnetic vacuum state. Nature 568:202–206. ISSN: 1476-4687
7. Bushunov AA et al (2020) Fabrication of antireflection microstructures on the surface of GaSe crystal by single-pulse femtosecond laser ablation. Opt Lett 45:5994–5997. ISSN: 1539-4794
8. Hagner M et al (2021) Ultrabroadband suppression of mid-infrared reflection losses of a layered semiconductor by nanopatterning with a focused ion beam. Opt Exp 29:33632. ISSN: 1094-4087 (2021)
9. Pupeza I et al (2020) Field-resolved infrared spectroscopy of biological systems. Nature 577:52–59. ISSN: 1476-4687
10. Süss B, Ringleb F, Heberle J (2016) New ultrarapid-scanning interferometer for FT-IR spectroscopy with microsecond time-resolution. Rev Sci Instrum 87:063113. ISSN: 1089-7623
11. Weigel A et al (2021) Attosecond-precision dual-oscillator infrared field-resolved spectroscopy employing electro-optic delay tracking in 2021 conference on lasers and electro-optics Europe & European quantum electronics conference (CLEO/Europe-EQEC). IEEE, vol 1. ISBN: 978-1-6654-1876-8
12. Baker MJ et al (2014) Using Fourier transform IR spectroscopy to analyze biological materials. Nat Protoc 9:1771–1791
13. Seidel M et al (2018) Multi-watt, multi-octave, mid-infrared femtosecond source. Sci Adv 4:eaaq1526
14. Wang Q et al (2019) Highly efficient broadband Mid-infrared generation (2.8–12.5 μm) based on a compact Cr:ZnS lasers and electro-optics Europe and European quantum electronics conference (CLEO/Europe-EQEC). IEEE, Piscataway
15. Vasilyev S et al (2020) Multi-octave infrared femtosecond continuum generation in Cr:ZnSGaSe and Cr:ZnS-ZGP tandems. In: Schunemann PG, Schepler KL (eds) Nonlinear frequency generation and conversion: materials and devices, vol XIX. SPIE, Bellingham, Washington
16. Degert J, Cornet M, Abraham E, Freysz E (2016) Simple and distortion-free optical ampling of terahertz pulses via heterodyne detection schemes. JOSA B 33:2045–2050. ISSN: 1520-8540

17. Aseev O et al (2019) High-precision ethanol measurement by mid-IR laser absorption spectroscopy for metrological applications. Opt Exp 27:5314–5325
18. Wallace WE (2021) In: Linstrom PJ, Mallard WG (eds) NIST chemistry WebBook, NIST standard reference database number 69. National Institute of Standards and Technology, Gaithersburg, Retrieved 15 May 2021
19. Muraviev AV, Smolski VO, Loparo ZE, Vodopyanov KL (2018) Massively parallel sensing of trace molecules and their isotopologues with broadband subharmonic mid-infrared frequency combs. Nat Photon 12:209–214. ISSN: 1749-4885

Appendix A
Data Archiving

The experimental raw data and simulation codes used in this thesis can be found on the data archive server of the Laboratory for Attosecond Physics at the Max Planck Institute of Quantum Optics:

//afs/ipp-garching.mpg.de/mpq/lap/publication-archive/theses/2021/Hofer, Christina (PhD).

For each figure, all necessary raw data, evaluation and plotting files are collected in a separate folder with the corresponding name. The data processing is described in the PDF '*DissertationDataArchiving.pdf*' and the final figures are collected in the corresponding folder.

© The Editor(s) (if applicable) and The Author(s), under exclusive license to Springer Nature Switzerland AG 2022

C. Hofer, *Detection Efficiency and Bandwidth Optimized Electro-Optic Sampling of Mid-Infrared Waves*, Springer Theses,
https://doi.org/10.1007/978-3-031-15328-0

Appendix B
Setup of the 2-μm EOS

See Fig. B.1.

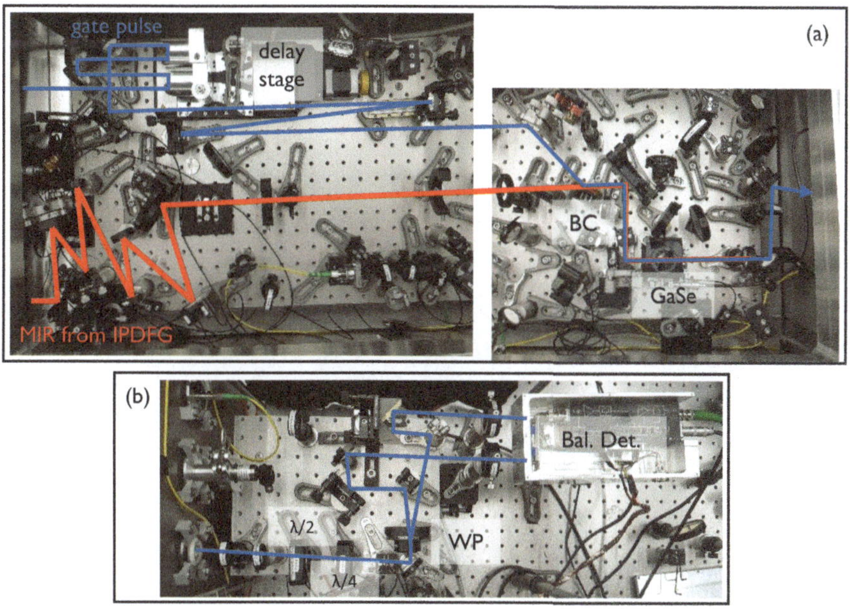

Fig. B.1 Picture of the experimental setup of the 2-μm EOS. **a** The gate pulses are variably delay with a motorized stage with two retroreflectors on top. MIR and gate beam are recombined using a ZnSe plate as a recombiner (BC) and focused into the GaSe crystal. Up to this point, the setup can be kept in vacuum environment. **b** Ellipsometry setup outside the vacuum chamber. The gate beam and generated sum-frequency radiation are transmitted through several filters and the waveplates. They are split in two beams for balanced detection (Bal. Det.) using a Wollaston Prism (WP)

© The Editor(s) (if applicable) and The Author(s), under exclusive license to Springer Nature Switzerland AG 2022

C. Hofer, *Detection Efficiency and Bandwidth Optimized Electro-Optic Sampling of Mid-Infrared Waves*, Springer Theses,
https://doi.org/10.1007/978-3-031-15328-0

Appendix C
Curriculum Vitae

Personal Information

Name: Christina Hofer
Date/Place of birth: June 23rd, 1993/Linz, Austria
Nationality: Austrian
E-mail: christina.hofer@mpq.mpg.de

Education

- **2016–2021**: PhD in Physics at the Max Planck Institute of Quantum Optics/LMU Munich
  Thesis title: ***Detection Efficiency and Bandwidth Optimized Electro-optic Sampling of Mid-Infrared Waves***
  Thesis advisor: Prof. Ferenc Krausz
  Graduated 'with greatest distinction'
  PhD representative
- **2014–2016**: Master of Science (Condensed Matter Physics), TU Munich
  Thesis title: ***Ultrabroadband Enhancement Cavities for Nonlinear Optics***
  Thesis advisor: Dr. Ioachim Pupeza
- **2011–2014**: Bachelor of Science (Physics), University of Innsbruck, Austria
  Thesis title: ***Quantensimulation der Spin-Spin Wechselwirkung geladener Ionen***
  Thesis advisor: Dr. Christian Roos

Honors and Awards

- 2022: Springer Thesis award for PhD thesis
- 2018: Stipend of the International Max Planck Research School on Advanced Photon Science (IMPRS-APS)
- 2015: 'Deutschlandstipendium' for Master's studies
- 2011-2014: Performance scholarship during Bachelor studies

© The Editor(s) (if applicable) and The Author(s), under exclusive license to Springer Nature Switzerland AG 2022

C. Hofer, *Detection Efficiency and Bandwidth Optimized Electro-Optic Sampling of Mid-Infrared Waves*, Springer Theses,
https://doi.org/10.1007/978-3-031-15328-0

List of Publications

Peer-Reviewed Journal Articles (co-author)

N. Lilienfein, C. Hofer, S. Holzberger, C. Matzer, P. Zimmermann, M. K. Trubetskov, V. Pervak, I. Pupeza, Enhancement cavities for few-cycle pulses, Optics Letters (2017): *During my Master's thesis, I developed characterization methods for laser pulses circulating inside an enhancement cavity.*

J. Xu, B. Globisch, C. Hofer, N. Lilienfein, T. P. Butler, N. Karpowicz, I. Pupeza, Three-octave terahertz pulses from optical rectification of 20-fs, 1 − μm, 78-MHz pulses in GaP, Journal of Physics B (2018): *I planned the electro-optic sampling setup for the THz pulse characterization in Gallium Phosphide and implemented it together with J. Xu.*

N. Lilienfein, C. Hofer, M. Högner, T. Saule, M. K. Trubetskov, V. Pervak, E. Fill, C. Riek, A. Leitenstorfer, J. Limpert, F. Krausz, I. Pupeza, Temporal solitions in free-space femtosecond enhancement cavities, Nature Photonics (2019): *During my Master's thesis, I did the first proof-of-principle experiments and developed a simple theoretical model for the soliton build-up.*

T. P. Butler, D. Gerz, C. Hofer, J. Xu, C. Gaida, T. Heuermann, M. Gebhardt, L. Vamos, W. Schweinberger, J. A. Gessner, T. Siefke, M. Heusinger, U. Zeitner, A. Apolonskiy, N. Karpowicz, J. Limpert, F, Krausz, I. Pupeza, Watt-scale 50-MHz source of single-cycle waveform-stable pulses in the molecular fingerprint region, Optics Letters (2019): *I planned and built the electro-optic sampling (EOS) setup, that uses gate pulses centered at* 2 μm *wavelength for the first time. I also developed the theoretical model and performed the electric-field reconstruction from the EOS measurements.*

B.-H. Chen, C. Hofer, I. Pupeza, P. Baum, Second-harmonic generation and self-phase modulation of few-cycle mid-infrared pulses, Optics Letters (2019): *I helped with the theoretical model and supported the experimental setup.*

T. P. Butler, N. Lilienfein, J. Xu, N. Nagl, C. Hofer, D. Gerz, K. F. Mak, C. Gaida, T. Heuermann, M. Gebhardt, J. Limpert, F. Krausz, I. Pupeza, Multi-octave spanning, Watt-level ultrafast mid-infrared source, Journal of Physics: Photonics (2019): *I developed the mid-infrared characterization setup and helped with the setup and characterization of the other channels.*

I. Pupeza, M. Huber, M. Trubetskov, W. Schweinberger, S. A. Hussain, C. Hofer, K. Fritsch, M. Poetzlberger, L. Vamos, E. Fill, T. V. Amotchkina, K. V. Kepesidis, A. Apolonskiy, N. Karpowicz, V. Pervak, O. Pronin, F. Fleischmann, A. Azzeer, M. Zigman, F. Krausz, Field-resolved infrared spectroscopy of biological systems, Nature (2020): *I planned, built, optimized and characterized the sensitivity of the EOS setup. I prepared several figures and co-wrote the supplementary material.*

M. Huber, M. K. Trubetskov, S. A. Hussain, W. Schweinberger, C. Hofer and I. Pupeza, Optimum sample thickness for trace analyte detection with field-resolved infrared spectroscopy, Analytical Chemistry (2020): *I planned, built, optimized and characterized the EOS detection.*

Q. Bournet, F. Guichard, M. Natile, Y. Zaouter, M. Joffre, A. Bonvalet, I. Pupeza, C. Hofer, F. Druon, M. Hanna, P. Georges, Enhanced intrapulse difference frequency generation in the mid-IR by spectrally-dependent polarization state, Optics Letters (2021): *Supported the design of the EOS setup.*

P. Steinleitner, N. Nagl, M. Kowalczyk, J. Zhang, V. Pervak, C. Hofer, A. Hudzikowski, J. Sotor, A. Weigel, F. Krausz, K. F. Mak, Single-cycle infrared waveform control, Nature Photonics (2022): *Supported the design, setup and (theoretical) understanding of the electro-optic detection.*

P. Sulzer, M. Högner, A.-K. Raab, L. Fürst, E. Fill, D. Gerz, C. Hofer, L. V. Voronina, I. Pupeza, Cavity-enhanced field-resolved spectroscopy, Nature Photonics (2022): Together with D. Gerz, I provided and operated the amplifier system, mid-IR generation and EOS detection.

Conference Talks

C.Hofer et al., Sensitivity of optical field measurements via electro-optic sampling, Frontiers of Nonlinear Physics (2019) [Invited talk].

C. Hofer et al., Mid-infrared electric field sampling approaching single-photon sensitivity, EPJ Web of Conferences, EUROPHOTON (2020) [Upgrade to invited talk].

C. Hofer et al., Electro-optic sampling with percent-level detection efficiency, CLEO Europe (2021).